LORENZO RAUSA

CNC
FANUC TURNING CYCLES

Description of the Parameters and Programming Examples

Copyright © cnc webschool ™ 2021
New York (U.S.A.)

info@cncwebschool.com
www.cncwebschool.com

All rights reserved under national law and international conventions.

First edition: January 2021 (R1.11).

ISBN: 9781727339888

Table of contents

Foreword

1. **List of functions** 11
   1.1 Description of G-code systems 11
   1.2 List of functions and functional groups 12
       1.2.1 From the function G0 to the function G26 12
       1.2.2 From the function G27 to the function G42.1 13
       1.2.3 From the function G43 to the function G66 14
       1.2.4 From the function G66.1 to the function G94 15
       1.2.5 From the function G91.1 to the function G99 16
2. **Introduction to Fanuc canned cycles** 17
   2.1 Canned cycles summary table 17
       2.1.1 Functions listed in ascending order 17
   2.2 Classification of canned cycles 19
       2.2.1 Non-repetitive cycles 19
       2.2.2 Repetitive cycles 19
       2.2.3 Grooving cycles 19
       2.2.4 Drilling, tapping, reaming or boring cycles 20
       2.2.5 Syntax used in the programming examples 20
   2.3 Summary table of cycles grouped by type 21
       2.3.1 Non-repetitive cycles 21
       2.3.2 Repetitive cycles 21
       2.3.3 Grooving cycles 22
       2.3.4 Drilling, tapping, boring or reaming cycles 22
3. **G77: single cut turning cycle** 23
   3.1 Description 23
   3.2 Cycle cancelation functions 24
   3.3 Cycle parameters 24
   3.4 Programming examples 26
       3.4.1 Cylindrical turning 26
       3.4.2 Taper turning 27
4. **G78: single cut threading cycle** 29

	4.1	Description 29
	4.2	Cycle cancelation functions 30
	4.3	Cycle-related NC parameters 30
	4.4	Cycle parameters.. 31
	4.5	Programming examples.. 34
		4.5.1 Cylindrical thread 34
		4.5.2 Tapered thread... 35
5.	**G79: single cut facing cycle** **37**
	5.1	Description 37
	5.2	Cycle cancelation functions 38
	5.3	Cycle parameters... 38
	5.4	Programming examples.. 41
		5.4.1 Straight face... 41
		5.4.2 Sloped shoulder 42
6.	**G70: finishing cycle or profile repetition**................................ **43**
	6.1	Description 43
	6.2	Cycle cancelation functions 44
	6.3	Cycle parameters... 44
	6.4	Programming examples.. 45
		6.4.1 Finishing 45
		6.4.2 Groove repetition.. 47
7.	**G71: roughing cycle along the Z-axis** **49**
	7.1	Description 49
		7.1.1 Roughing cycle type 1 50
		7.1.2 Roughing cycle type 2 51
		7.1.3 Setting of the approaching feed rate 52
	7.2	Cycle cancelation functions 52
	7.3	Cycle-related NC parameters 52
	7.4	Cycle parameters... 53
	7.5	Programming examples.. 55
		7.5.1 External rough turning 55
		7.5.2 Internal rough turning 57
8.	**G72: roughing cycle along the X-axis**....................................... **59**
	8.1	Description 59
		8.1.1 Roughing cycle type 1 60
		8.1.2 Roughing cycle type 2 61
		8.1.3 Setting of the approaching feed rate 62
	8.2	Cycle cancelation functions 62

	8.3	Cycle-related NC parameters 62
	8.4	Cycle parameters.. 63
	8.5	Programming example 65
		8.5.1 Roughing.. 65
9.	**G73: pattern repeating cycle**.. 67
	9.1	Description 67
	9.2	Cycle cancelation functions 68
	9.3	Cycle parameters.. 68
	9.4	Programming example 70
		9.4.1 Roughing.. 70
10.	**G76: multi-pass threading cycle** 73
	10.1	Description 73
	10.2	Cycle cancelation functions 74
	10.3	Cycle parameters.. 74
	10.4	Programming examples 80
		10.4.1 Cylindrical thread 80
		10.4.2 Tapered thread.. 81
11.	**G74: grooving cycle along the Z-axis** 83
	11.1	Description 83
	11.2	Cycle cancelation functions 84
	11.3	Cycle parameters.. 84
	11.4	Programming examples 86
		11.4.1 Front groove 86
		11.4.2 External turning with chip breaking 87
		11.4.3 Drilling with chip breaking 88
		11.4.4 Boring with chip breaking................................ 89
12.	**G75: grooving cycle along the X-axis**................................... 91
	12.1	Description 91
	12.2	Cycle cancelation functions 92
	12.3	Cycle parameters.. 92
	12.4	Programming example 94
		12.4.1 Radial groove 94
		12.4.2 Parting-off with chip breaking......................... 95
		12.4.3 Facing with chip breaking................................ 96
		12.4.4 Radial drilling with chip breaking 97
13.	**G83: drilling cycle along the Z-axis**...................................... 99
	13.1	Description 99
	13.2	Cycle-related NC parameters 102

- 13.3 Cycle cancelation functions 102
- 13.4 Cycle parameters 103
- 13.5 Programming example 106
  - 13.5.1 Axial drilling 106
- **14. G87: drilling cycle along the X-axis 107**
  - 14.1 Description 107
  - 14.2 Cycle-related NC parameters 109
  - 14.3 Cycle cancelation functions 109
  - 14.4 Cycle parameters 110
  - 14.5 Programming examples 113
    - 14.5.1 Radial drilling 113
    - 14.5.2 Radial drilling of three holes 114
    - 14.5.3 Parting-off 116
    - 14.5.4 Three radial grooves 117
- **15. G84: tapping cycle along the Z-axis 119**
  - 15.1 Description 119
  - 15.2 Cycle-related NC parameters 121
  - 15.3 Axial rigid peck tapping 121
  - 15.4 Cycle cancelation functions 122
  - 15.5 Cycle parameters 122
  - 15.6 Programming example 125
    - 15.6.1 Axial tapping 125
- **16. G88: tapping cycle along the X-axis 127**
  - 16.1 Description 127
  - 16.2 Cycle-related NC parameters 129
  - 16.3 Axial rigid peck tapping 129
  - 16.4 Cycle cancelation functions 130
  - 16.5 Cycle parameters 130
  - 16.6 Programming example 132
    - 16.6.1 Radial tapping 132
- **17. G85: boring cycle along the Z-axis 135**
  - 17.1 Description 135
  - 17.2 Cycle cancelation functions 137
  - 17.3 Cycle parameters 137
  - 17.4 Programming example 140
    - 17.4.1 Axial boring 140
- **18. G89: boring cycle along the X-axis 141**

18.1	Description	141
18.2	Cycle cancelation functions	143
18.3	Cycle parameters	143
18.4	Programming example	145
	18.4.1 Radial reaming	145

Foreword

The purpose of this book is to explain the Fanuc turning canned cycles through a new didactic concept.

In different manuals it is easy to find contrasting descriptions regarding the Fanuc turning canned cycles.
Some manuals present the G74 function as an axial drilling cycle and others present it as a grooving cycle along the Z-axis.
The G75 function is also described in some texts as a radial grooving cycle, while in others it is defined as a radial drilling cycle.
It should be added that the G75 function is also able to perform a facing cut with chip breaking.

The book aims to explain the Fanuc turning cycles in a definite way by adopting a new didactic method that is not limited to the simple description of cycle parameters, but includes all the machining operations that each cycle is able to perform.

Lorenzo Rausa

cnc webschool

1. List of functions

1.1 Description of G-code systems

Fanuc manuals list the programming codes in three different columns: "A", "B" and "C".
These letters identify three different groups of codes according to which the same functions can be enabled by different "G" codes (e.g. in the "B" code system, function G95 sets the cutting feed rate in mm/rev, while in the "A" code system the cutting feed rate is set by G99).
Originally, the G code systems identified the geographical area where the machine was manufactured. Asian manufacturers used the "A-code" system; those from Europe used the "B-code" system and those from America used the "C-code" system.

G code system			Group	Function
A	B	C		
G00	G00	G00		Positioning (Rapid traverse)
G01	G01	G01		Linear interpolation (Cutting feed)
G02	G02	G02		Circular interpolation CW or helical interpolation CW

Nowadays, it is still possible to find European and American manufacturers who use the "A-code" system, as it is more widespread and well known.

This feature of Fanuc's numerical controls is the first thing to clarify and check when the programmer starts running a new machine.

In the official Fanuc manuals, functions are outlined according to the "A-code" system. In this book, functions are outlined according to the European "B-code" system.
In the next paragraphs, you will find the tables of equivalence that contain the code definitions according to the system they belong to.
The column "Group" specifies the group to which each function belongs. Remember that modal functions belonging to the same group overwrite each other.

1.2 List of functions and functional groups

Below is the list of functions as reported in the official Fanuc manuals.

1.2.1 From the function G0 to the function G26

G code system A	G code system B	G code system C	Group	Function
G00	G00	G00	01	Positioning (Rapid traverse)
G01	G01	G01		Linear interpolation (Cutting feed)
G02	G02	G02		Circular interpolation CW or helical interpolation CW
G03	G03	G03		Circular interpolation CCW or helical interpolation CCW
G02.2	G02.2	G02.2		Involute interpolation CW
G02.3	G02.3	G02.3		Exponential interpolation CW
G02.4	G02.4	G02.4		3-dimensional coordinate system conversion CW
G03.2	G03.2	G03.2		Involute interpolation CCW
G03.3	G03.3	G03.3		Exponential interpolation CCW
G03.4	G03.4	G03.4		3-dimensional coordinate system conversion CCW
G04	G04	G04	00	Dwell
G04.1	G04.1	G04.1		G code preventing buffering
G05	G05	G05		AI contour control (command compatible with high precision contour control), High-speed cycle machining, High-speed binary program operation
G05.1	G05.1	G05.1		AI contour control / Nano smoothing / Smooth interpolation
G05.4	G05.4	G05.4		HRV3, 4 on/off
G06.2	G06.2	G06.2	01	NURBS interpolation
G07	G07	G07	00	Hypothetical axis interpolation
G07.1 (G107)	G07.1 (G107)	G07.1 (G107)		Cylindrical interpolation
G08	G08	G08		AI contour control (advanced preview control compatible command)
G09	G09	G09		Exact stop
G10	G10	G10		Programmable data input
G10.6	G10.6	G10.6		Tool retract and recover
G10.9	G10.9	G10.9		Programmable switching of diameter/radius specification
G11	G11	G11		Programmable data input mode cancel
G12.1 (G112)	G12.1 (G112)	G12.1 (G112)	21	Polar coordinate interpolation mode
G13.1 (G113)	G13.1 (G113)	G13.1 (G113)		Polar coordinate interpolation cancel mode
G17	G17	G17	16	XpYp plane selection
G17.1	G17.1	G17.1		Plane conversion function
G18	G18	G18		ZpXp plane selection
G19	G19	G19		YpZp plane selection
G20	G20	G70	06	Input in inch
G21	G21	G71		Input in mm
G22	G22	G22	09	Stored stroke check function on
G23	G23	G23		Stored stroke check function off
G25	G25	G25	08	Spindle speed fluctuation detection off
G26	G26	G26		Spindle speed fluctuation detection on

Fig. 1 List of Fanuc functions from G0 to G26

1.2.2 From the function G27 to the function G42.1

G code system			Group	Function
A	B	C		
G27	G27	G27	00	Reference position return check
G28	G28	G28		Return to reference position
G28.2	G28.2	G28.2		In-position check disable reference position return
G29	G29	G29		Movement from reference position
G30	G30	G30		2nd, 3rd and 4th reference position return
G30.1	G30.1	G30.1		Floating reference point return
G30.2	G30.2	G30.2		In-position check disable 2nd, 3rd, or 4th reference position return
G31	G31	G31		Skip function
G31.8	G31.8	G31.8		EGB-axis skip
G32	G33	G33	01	Threading
G34	G34	G34		Variable lead threading
G35	G35	G35		Circular threading CW
G36	G36	G36		Circular threading CCW (When bit 3 (G36) of parameter No. 3405 is set to 1) or Automatic tool offset (X axis) (When bit 3 (G36) of parameter No. 3405 is set to 0)
G37	G37	G37		Automatic tool offset (Z axis) (When bit 3 (G36) of parameter No. 3405 is set to 0)
G37.1	G37.1	G37.1		Automatic tool offset (X axis) (When bit 3 (G36) of parameter No. 3405 is set to 1)
G37.2	G37.2	G37.2		Automatic tool offset (Z axis) (When bit 3 (G36) of parameter No. 3405 is set to 1)
G38	G38	G38		Tool radius/tool nose radius compensation: with vector held
G39	G39	G39		Tool radius/tool nose radius compensation: corner rounding interpolation
G40	G40	G40	07	Tool radius/tool nose radius compensation : cancel
G41	G41	G41		Tool radius/tool nose radius compensation : left
G42	G42	G42		Tool radius/tool nose radius compensation : right
G41.2	G41.2	G41.2		3-dimensional cutter compensation : left (type 1)
G41.3	G41.3	G41.3		3-dimensional cutter compensation : (leading edge offset)
G41.4	G41.4	G41.4		3-dimensional cutter compensation : left (type 1) (FS16i-compatible command)
G41.5	G41.5	G41.5		3-dimensional cutter compensation : left (type 1) (FS16i-compatible command)
G41.6	G41.6	G41.6		3-dimensional cutter compensation : left (type 2)
G42.2	G42.2	G42.2		3-dimensional cutter compensation : right (type 1)
G42.4	G42.4	G42.4		3-dimensional cutter compensation : right (type 1) (FS16i-compatible command)
G42.5	G42.5	G42.5		3-dimensional cutter compensation : right (type 1) (FS16i-compatible command)
G42.6	G42.6	G42.6		3-dimensional cutter compensation : right (type 2)
G40.1	G40.1	G40.1	19	Normal direction control cancel mode
G41.1	G41.1	G41.1		Normal direction control left on
G42.1	G42.1	G42.1		Normal direction control right on

Fig. 2. List of Fanuc functions from G27 to G42.1

1.2.3 From the function G43 to the function G66

G code system A	G code system B	G code system C	Group	Function
G43	G43	G43	23	Tool length compensation + (Bit 3 (TCT) of parameter No. 5040 must be "1".)
G44	G44	G44		Tool length compensation - (Bit 3 (TCT) of parameter No. 5040 must be "1".)
G43.4	G43.4	G43.4		Tool center point control (type 1) (Bit 3 (TCT) of parameter No. 5040 must be "1".)
G43.5	G43.5	G43.5		Tool center point control (type 2) (Bit 3 (TCT) of parameter No. 5040 must be "1".)
G43.7 (G44.7)	G43.7 (G44.7)	G43.7 (G44.7)		Tool offset (Bit 3 (TCT) of parameter No. 5040 must be "1".)
G44.1	G44.1	G44.1		Tool offset conversion (Bit 3 (TCT) of parameter No. 5040 must be "1".)
G49 (G49.1)	G49 (G49.1)	G49 (G49.1)		Tool length compensation cancel (Bit 3 (TCT) of parameter No. 5040 must be "1".)
G50	G92	G92	00	Coordinate system setting or max spindle speed clamp
G50.3	G92.1	G92.1		Workpiece coordinate system preset
-	G50	G50	18	Scaling cancel
-	G51	G51		Scaling
G50.1	G50.1	G50.1	22	Programmable mirror image cancel
G51.1	G51.1	G51.1		Programmable mirror image
G50.2 (G250)	G50.2 (G250)	G50.2 (G250)	20	Polygon turning cancel
G51.2 (G251)	G51.2 (G251)	G51.2 (G251)		Polygon turning
G50.4	G50.4	G50.4	00	Cancel synchronous control
G50.5	G50.5	G50.5		Cancel composite control
G50.6	G50.6	G50.6		Cancel superimposed control
G51.4	G51.4	G51.4		Start synchronous control
G51.5	G51.5	G51.5		Start composite control
G51.6	G51.6	G51.6		Start superimposed control
G52	G52	G52		Local coordinate system setting
G53	G53	G53		Machine coordinate system setting
G53.1	G53.1	G53.1		Tool axis direction control
G53.6	G53.6	G53.6		Tool center point retention type tool axis direction control
G54 (G54.1)	G54 (G54.1)	G54 (G54.1)	14	Workpiece coordinate system 1 selection
G55	G55	G55		Workpiece coordinate system 2 selection
G56	G56	G56		Workpiece coordinate system 3 selection
G57	G57	G57		Workpiece coordinate system 4 selection
G58	G58	G58		Workpiece coordinate system 5 selection
G59	G59	G59		Workpiece coordinate system 6 selection
G54.4	G54.4	G54.4	26	Workpiece setting error compensation
G60	G60	G60	00	Single direction positioning
G61	G61	G61	15	Exact stop mode
G62	G62	G62		Automatic corner override mode
G63	G63	G63		Tapping mode
G64	G64	G64		Cutting mode
G65	G65	G65	00	Macro call
G66	G66	G66		Macro modal call A

Fig. 3. List of Fanuc functions from G43 to G66

1.2.4 From the function G66.1 to the function G94

G code system A	G code system B	G code system C	Group	Function
G66.1	G66.1	G66.1	12	Macro modal call B
G67	G67	G67		Macro modal call A/B cancel
G68	G68	G68	04	Mirror image on for double turret or balance cutting mode
G68.1	G68.1	G68.1	17	Coordinate system rotation start or 3-dimensional coordinate system conversion mode on
G68.2	G68.2	G68.2		Tilted working plane indexing command
G68.3	G68.3	G68.3		Tilted working plane indexing command by tool axis direction
G68.4	G68.4	G68.4		Tilted working plane indexing command (incremental multi-command)
G69	G69	G69	04	Mirror image off for double turret or balance cutting mode cancel
G69.1	G69.1	G69.1	17	Coordinate system rotation cancel or 3-dimensional coordinate system conversion mode off
G70	G70	G72	00	Finishing cycle
G71	G71	G73		Stock removal in turning
G72	G72	G74		Stock removal in facing
G73	G73	G75		Pattern repeating cycle
G74	G74	G76		End face peck drilling cycle
G75	G75	G77		Outer diameter/internal diameter drilling cycle
G76	G76	G78		Multiple-thread cutting cycle
G71	G71	G72	01	Traverse grinding cycle
G72	G72	G73		Traverse direct sizing/grinding cycle
G73	G73	G74		Oscillation grinding cycle
G74	G74	G75		Oscillation direct sizing/grinding cycle
G80	G80	G80	10	Canned cycle cancel for drilling
				Electronic gear box : synchronization cancellation
G81.1	G81.1	G81.1	00	Chopping function/High precision oscillation function
G80.4	G80.4	G80.4	28	Electronic gear box: synchronization cancellation
G81.4	G81.4	G81.4		Electronic gear box: synchronization start
G80.5	G80.5	G80.5	27	Electronic gear box 2 pair: synchronization cancellation
G81.5	G81.5	G81.5		Electronic gear box 2 pair: synchronization start
G81	G81	G81		Spot drilling (FS15-T format)
				Electronic gear box : synchronization start
G82	G82	G82		Counter boring (FS15-T format)
G83	G83	G83		Cycle for face drilling
G83.1	G83.1	G83.1		High-speed peck drilling cycle (FS15-T format)
G83.5	G83.5	G83.5		High-speed peck drilling cycle
G83.6	G83.6	G83.6		Peck drilling cycle
G84	G84	G84	10	Cycle for face tapping
G84.2	G84.2	G84.2		Rigid tapping cycle (FS15-T format)
G85	G85	G85		Cycle for face boring
G87	G87	G87		Cycle for side drilling
G87.5	G87.5	G87.5		High-speed peck drilling cycle
G87.6	G87.6	G87.6		Peck drilling cycle
G88	G88	G88		Cycle for side tapping
G89	G89	G89		Cycle for side boring
G90	G77	G20	01	Outer diameter/internal diameter cutting cycle
G92	G78	G21		Threading cycle
G94	G79	G24		End face turning cycle

Fig. 4. List of Fanuc functions from G66.1 to G94

1.2.5 From the function G91.1 to the function G99

G code system			Group	Function
A	B	C		
G91.1	G91.1	G91.1	00	Maximum specified incremental amount check
G96	G96	G96	02	Constant surface speed control
G97	G97	G97		Constant surface speed control cancel
G96.1	G96.1	G96.1	00	Spindle indexing execution (waiting for completion)
G96.2	G96.2	G96.2		Spindle indexing execution (not waiting for completion)
G96.3	G96.3	G96.3		Spindle indexing completion check
G96.4	G96.4	G96.4		SV speed control mode ON
G93	G93	G93	05	Inverse time feed
G98	G94	G94		Feed per minute
G99	G95	G95		Feed per revolution
-	G90	G90	03	Absolute programming
-	G91	G91		Incremental programming
-	G98	G98	11	Canned cycle : return to initial level
-	G99	G99		Canned cycle : return to R point level

Fig. 5. List of Fanuc functions from G91.1 to G99

2. Introduction to Fanuc canned cycles

2.1 Canned cycles summary table

2.1.1 Functions listed in ascending order

Function (B system)	Function (A system)	Function (C system)	Type of machining operation
G70	G70	G72	Finishing cycle or profile repetition
G71	G71	G73	Roughing cycle along Z-axis
G72	G72	G74	Roughing cycle along X-axis
G73	G73	G75	Pattern repeating cycle
G74	G74	G76	Grooving cycle along Z-axis
G75	G75	G77	Grooving cycle along X-axis
G76	G76	G78	Threading cycle
G77	G90	G20	Single cut turning cycle
G78	G92	G21	Single cut threading cycle
G79	G94	G24	Single cut facing cycle

G80	G80	G80	Canned cycle deactivation
G83	G83	G83	Drilling cycle along Z-axis
G84	G84	G84	Tapping cycle along Z-axis
G85	G85	G85	Boring or reaming cycle along Z-axis
G87	G87	G87	Drilling cycle along X-axis
G88	G88	G88	Tapping cycle along X-axis
G89	G89	G89	Boring or reaming cycle along X-axis

2.2 Classification of canned cycles

2.2.1 Non-repetitive cycles

Non-repetitive cycles are cycles that do not execute a complete machining operation but perform just a single cut. They involve a rapid approach to the workpiece, a single working cut, retraction from the workpiece and return to the starting position.

Fanuc numerical controls provide non-repetitive cycles to perform turning cuts (G77), facing cuts (G79), and threading cuts (G78).

These cycles belong to group 1. Therefore, you only need to program another function belonging to the same group (e.g. G0 or G1) to cancel them.

In this course, the finishing cycle (G70) has been included in the group of non-repetitive cycles because it does not repeat any type of operation.

2.2.2 Repetitive cycles

Repetitive cycles are cycles that execute a complete machining operation. They are called "repetitive" because they continue to repeat cutting passes until a specific machining operation is completed.

Fanuc numerical controls provide repetitive cycles for roughing operations along the Z-axis (G71), roughing operations along the X-axis (G72), roughing operations for profiled workpieces (G73) and for complete threading operations (G76).

These cycles are self-canceling and no further functions are necessary to deactivate them.

2.2.3 Grooving cycles

These cycles perform grooves both along the Z-axis (G74) and the X-axis (G75).

They cut in the direction of the groove with the possibility to program a chip break and widen the groove with passes parallel to the first one until the programmed value is reached. These cycles are also used for drilling and turning operations with chip breaking.

These cycles are self-canceling and no further functions are necessary to deactivate them.

2.2.4 Drilling, tapping, reaming or boring cycles

These cycles have been grouped as they are very similar. They are used for:
- Drilling along the Z-axis (G83) with chip removal or chip breaking.
- Drilling along the X-axis (G87) with chip removal or chip breaking.
- Tapping along the Z-axis (G84).
- Tapping along the X-axis (G88).
- Boring along the Z-axis (G85).
- Boring along the X-axis (G89).

These cycles are canceled with the function G80.

2.2.5 Syntax used in the programming examples

All the following programming examples for Fanuc turning cycles are operational in the training and graphic simulation software "Siemens Sinumerik Operate 4.4".

The machine template is called "Lathe with driven tool (ISO dialect)".
This machine is programmed in ISO language.

The Siemens programming syntax is used when the ISO language does not offer standard programming functions, common to all machines, such as those required for the definition of the stock dimensions or the selection of driven tools.

For this reason, in the programming examples, G290 function is used, when necessary, for the activation of the Siemens language, and G291 function for the activation of the Fanuc language.

To use the suggested programming examples on real machines, replace the part of the program between these functions with the specific program required by your machine.

2.3 Summary table of cycles grouped by type

2.3.1 Non-repetitive cycles

Function (B system)	Function (A system)	Function (C system)	Type of machining operation
G77	G90	G20	Single cut turning cycle
G79	G94	G24	Single cut facing cycle
G78	G92	G21	Single cut threading cycle
G70	G70	G72	Finishing cycle or profile repetition

2.3.2 Repetitive cycles

Function (B system)	Function (A system)	Function (C system)	Type of machining operation
G71	G71	G73	Roughing cycle along Z-axis
G72	G72	G74	Roughing cycle along X-axis
G73	G73	G75	Pattern repeating cycle
G76	G76	G78	Threading cycle

2.3.3 Grooving cycles

Function (B system)	Function (A system)	Function (C system)	Type of machining operation
G74	G74	G76	Grooving cycle along Z-axis
G75	G75	G77	Grooving cycle along X-axis

2.3.4 Drilling, tapping, boring or reaming cycles

Function (B system)	Function (A system)	Function (C system)	Type of machining operation
G83	G83	G83	Drilling cycle along Z-axis
G87	G87	G87	Drilling cycle along X-axis
G84	G84	G84	Tapping cycle along Z-axis
G88	G88	G88	Tapping cycle along X-axis
G85	G85	G85	Boring or reaming cycle along Z-axis
G89	G89	G89	Boring or reaming cycle along X-axis
G80	G80	G80	Cycle deactivation function

3. G77: single cut turning cycle
(G90-A, G20-C)

3.1 Description

This cycle performs a single turning cut, starting from the point where the tool is located before the cycle.

The cycle carries out four movements.

1. Rapid movement along the X-axis, from the point programmed before the cycle to the X-coordinate programmed in block G77.
2. Working movement along the Z-axis, from the current point to the Z-coordinate programmed in block G77.
3. Working movement from the end of the cut to the starting X-coordinate programmed before the cycle.
4. Rapid return movement to the starting Z-coordinate programmed before the cycle.

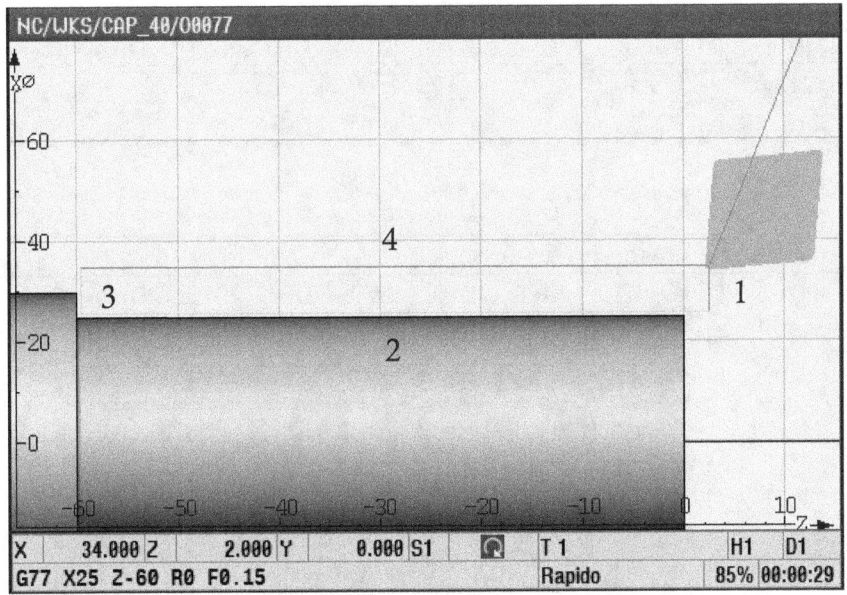

Fig. 6. G77: cycle movements

3.2 Cycle cancelation functions

Program a G-code of group 01 to cancel this cycle. If the machine uses the "A-code" system, the G-code must be different from G90, G92 or G94.

3.3 Cycle parameters

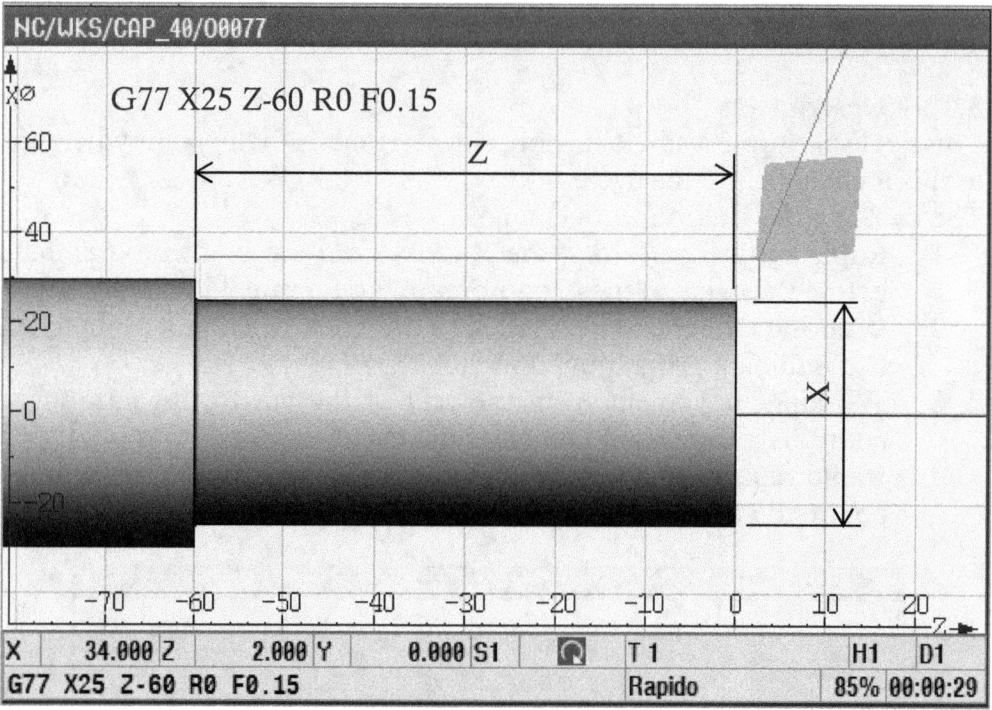

Fig. 7. G77: cycle parameters

Parameter	Description
X	Cutting diameter.
(U)	Use the letter "U" to set the cutting diameter as the diametrical distance along the X-axis between the cycle start point and the cutting diameter (with negative sign for external cuts).
Notes	For taper turning, it is necessary to program the arrival diameter (the starting point will be calculated by the cycle using the parameter "R").

Z (W)	Arrival point of the cut. The letter Z indicates the absolute coordinate (referring to the workpiece zero point), along the Z-axis, of the cutting end point. Use the letter "W" to set the cutting length as the distance along the Z-axis from the point programmed before the cycle to the arrival point of the cut.	
R	Taper value. It is equal to the starting diameter of the cut, minus the arrival diameter, divided by two. The starting and arrival diameters might not correspond to the taper diameters, but to the diameters of the cut to be performed, calculated on the basis of the tool positions in Z at the beginning and at the end of the cycle. When the starting diameter is smaller than the arrival diameter, the R value is negative (-). When the value is equal to zero or when the parameter is not programmed, the lathe turns parallel to the Z-axis.	
F	Working feed rate.	

3.4 Programming examples

3.4.1 Cylindrical turning

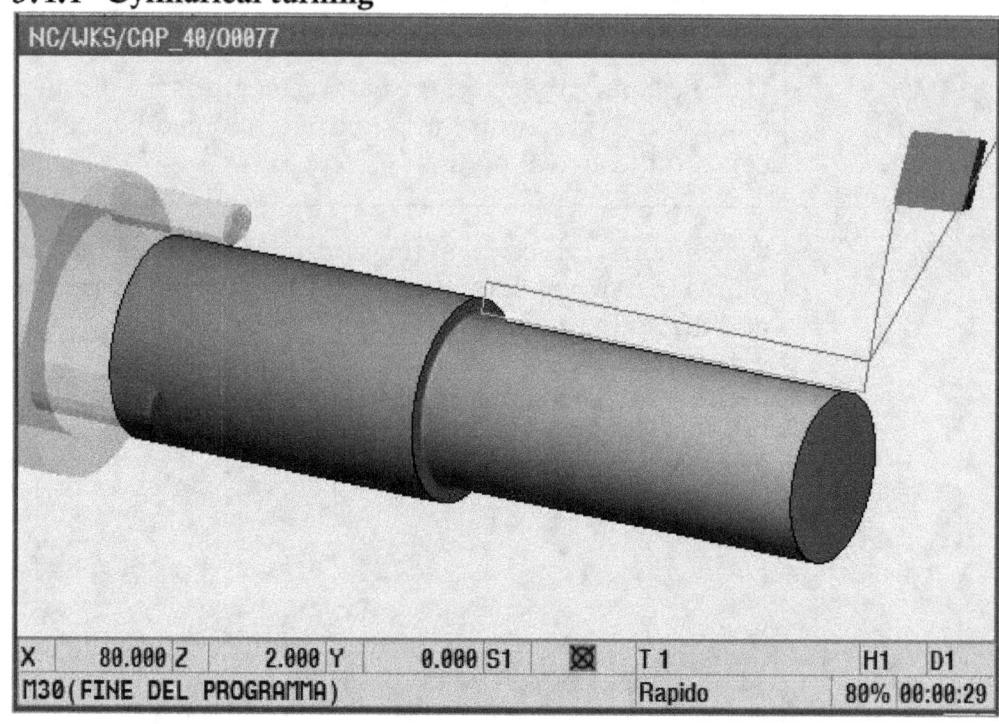

Fig. 8. G77: programming example

```
%
O0771
(G77 CYLINDRICAL TURNING)
G290 ;SIEMENS LANGUAGE ACTIVATION
WORKPIECE(,,,"CYLINDER",0,0,-110,-100,30)
G291 ;FANUC LANGUAGE ACTIVATION
G18 (X-Z PLANE)
G90 (ABSOLUTE PROGRAMMING)

T0101 (EXTERNAL TURNING TOOL)
G97 S1000 M4 (SPINDLE CONSTANT REVOLUTION)
G95 (WORKING FEED RATE IN MM/REV)

G0 X34 Z2 (CYCLE START POINT)
G77 X25 Z-60 R0 F0.15
G0 X80 (RETRACTION AND CYCLE CANCELATION)
M5 (STOP SPINDLE ROTATION)
M30 (END OF PROGRAM)
%
```

3.4.2 Taper turning

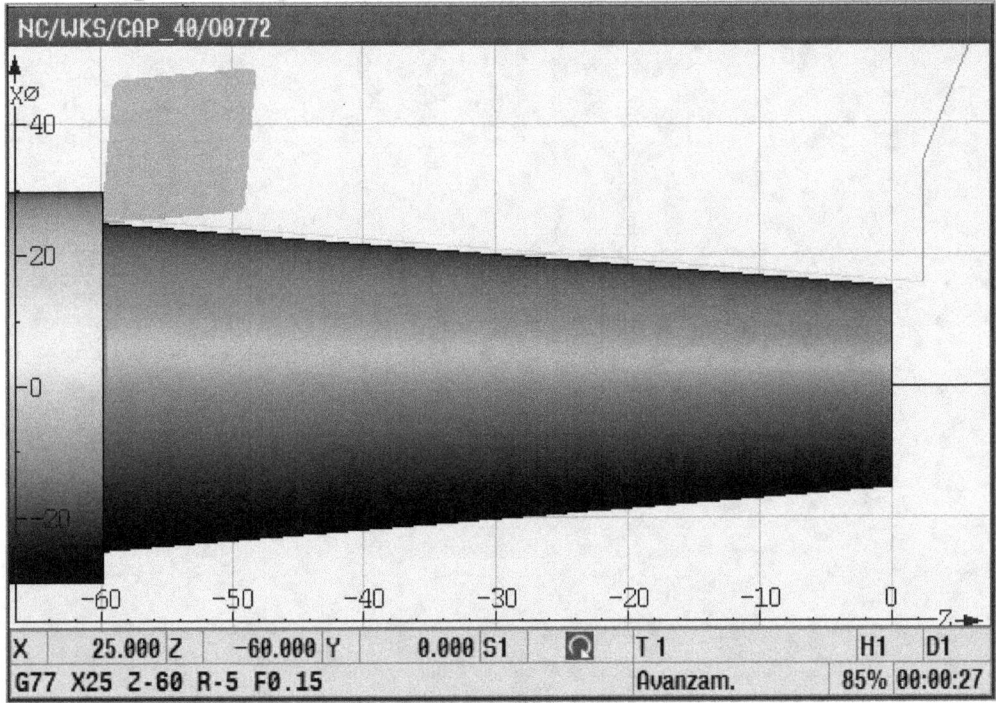

Fig. 9. G77: programming example

```
%
O0772
(G77 TAPER TURNING)
G290 ;SIEMENS LANGUAGE ACTIVATION
WORKPIECE(,,,"CYLINDER",0,0,-110,-100,30)
G291 ;FANUC LANGUAGE ACTIVATION
G18 (X-Z PLANE)
G90 (ABSOLUTE PROGRAMMING)
T0101 (EXTERNAL TURNING TOOL)
G97 S1000 M4 (SPINDLE CONSTANT REVOLUTION)
G95 (WORKING FEED RATE IN MM/REV)

G0 X34 Z2 (CYCLE START POINT)
G77 X25 Z-60 R-5 F0.15
G0 X80 (RETRACTION AND CYCLE CANCELATION)
M5 (STOP SPINDLE ROTATION)
M30(END OF PROGRAM)
%
```

In this example, X15 and Z2 are the coordinates of the starting point of the cut.

28

4. G78: single cut threading cycle
(G92-A, G21-C)

4.1 Description

This cycle performs a single threading cut, starting from the point where the tool is located before the cycle.

The cycle carries out four movements.

1. Rapid movement along the X-axis, from the point programmed before the cycle to the X-coordinate programmed in block G78.
2. Threading operation along the Z-axis, from the current point to the Z-coordinate programmed in block G78. The thread exit direction (45° or 90°) depends on the value set in the system parameters which will be explained later.
3. Rapid return movement to the starting X-coordinate programmed before the cycle.
4. Rapid return movement to the starting Z-coordinate programmed before the cycle.

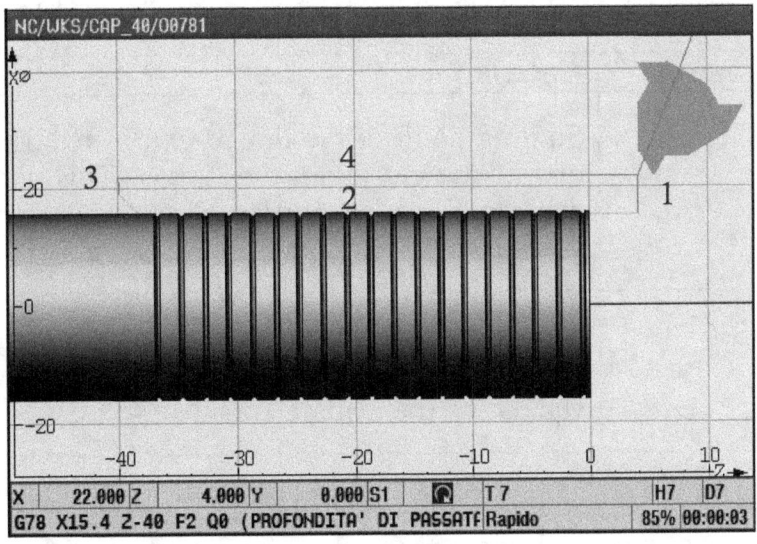

Fig. 10. G77: cycle movements

4.2 Cycle cancelation functions

Program a G-code of group 01 to cancel this cycle. If the machine uses the "A-code" system, the G-code must be different from G90, G92 or G94.

4.3 Cycle-related NC parameters

If there is a groove at the end of the thread, it is recommended to allow the tool to exit at a 90° angle. If there is no groove at the end of the thread, it is recommended to allow the tool to exit more gradually at an angle that can be set by using a parameter. The direction and starting point of the thread exit can be programmed by using the following NC parameters.

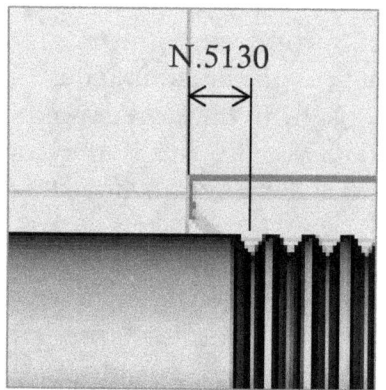 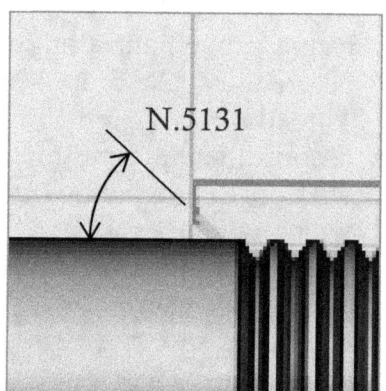

Fig. 11. These parameters define the thread exit direction.

Parameter	Description
N. 5130	Starting point of the thread exit in relation to the end of the cut along the Z-axis. This distance is expressed as a proportion of the lead with values between 0.1F and 12.7F in increments of 0.1.
N. 5131	Thread exit direction expressed as an angle with values between 1 and 89 degrees. If the value is equal to 0, the exit angle will be 45°.

The parameters specifying the starting chamfer distance and chamfer angle are the same in this cycle and in the G76 threading cycle.

4.4 Cycle parameters

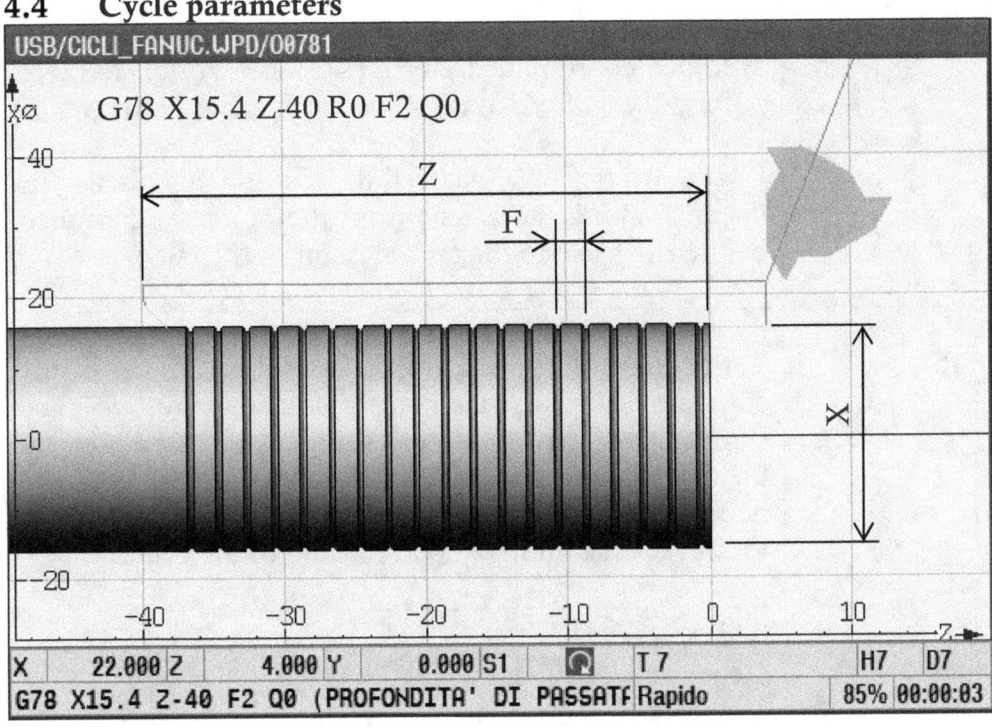

Fig. 12. G78: cycle parameters

Parameter	Description
X	Cutting diameter.
(U)	Use the letter "U" to set the cutting diameter as the diametrical distance along the X-axis between the cycle start point and the cutting diameter (with negative sign for external cuts).
Notes	For taper threading, it is necessary to program the arrival diameter (the starting point will be calculated by the cycle using the parameter "R").

Z (W)	Arrival point of the cut. The letter Z indicates the absolute coordinate (referring to the workpiece zero point), along the Z-axis, of the cutting end point. Use the letter "W" to set the cutting length as the distance along the Z-axis from the point programmed before the cycle to the arrival point of the cut.
R	Taper value. It is equal to the starting diameter of the cut, minus the arrival diameter, divided by two. The starting and arrival diameters might not correspond to the taper diameters, but to the diameters of the cut to be performed, calculated on the basis of the tool positions in Z at the beginning and at the end of the cycle. 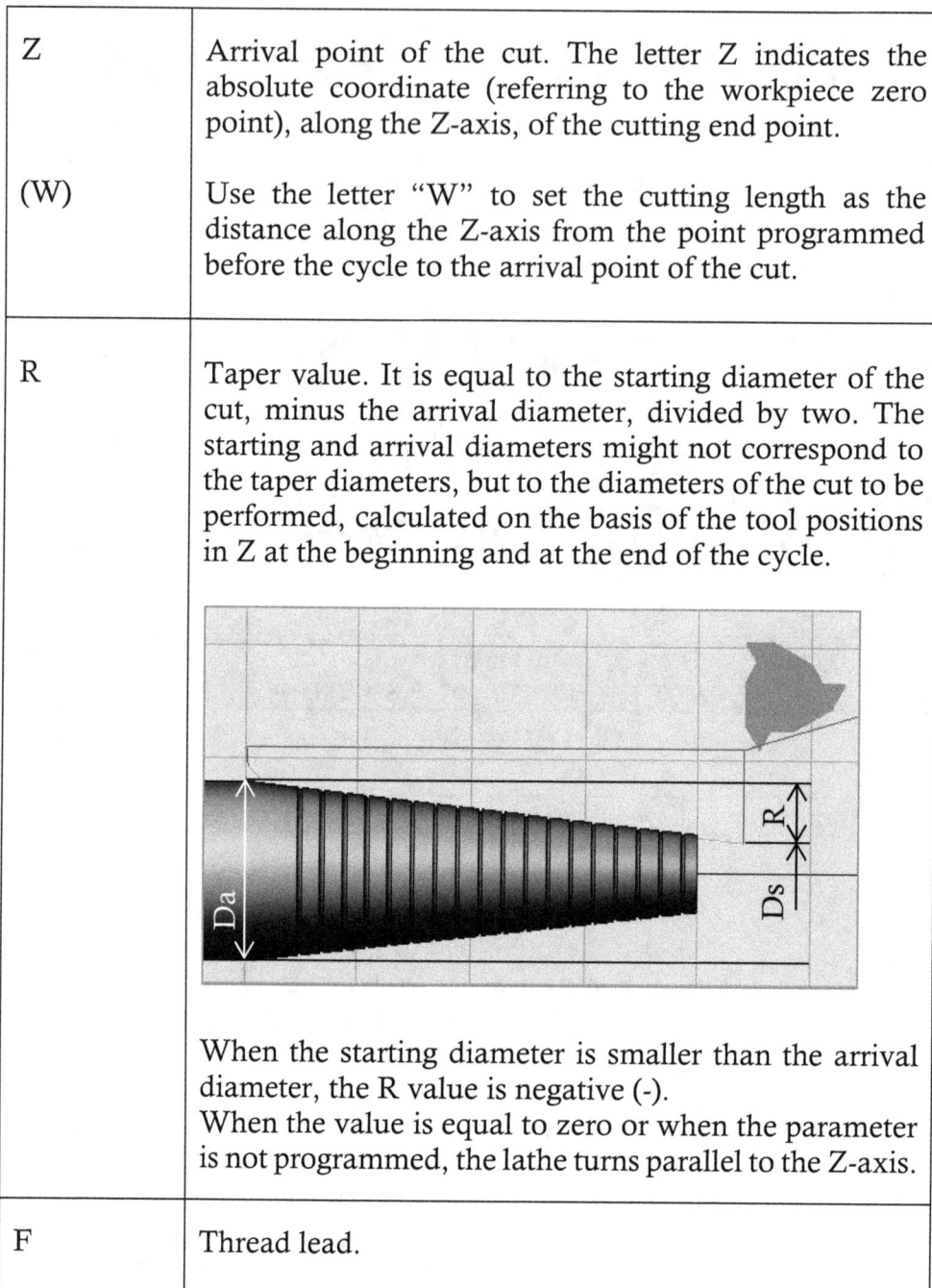 When the starting diameter is smaller than the arrival diameter, the R value is negative (-). When the value is equal to zero or when the parameter is not programmed, the lathe turns parallel to the Z-axis.
F	Thread lead.

Q	Starting angle of the threading cut. It is expressed in thousandths of a degree (i.e. an angle of 180° = Q180000).

4.5 Programming examples

4.5.1 Cylindrical thread

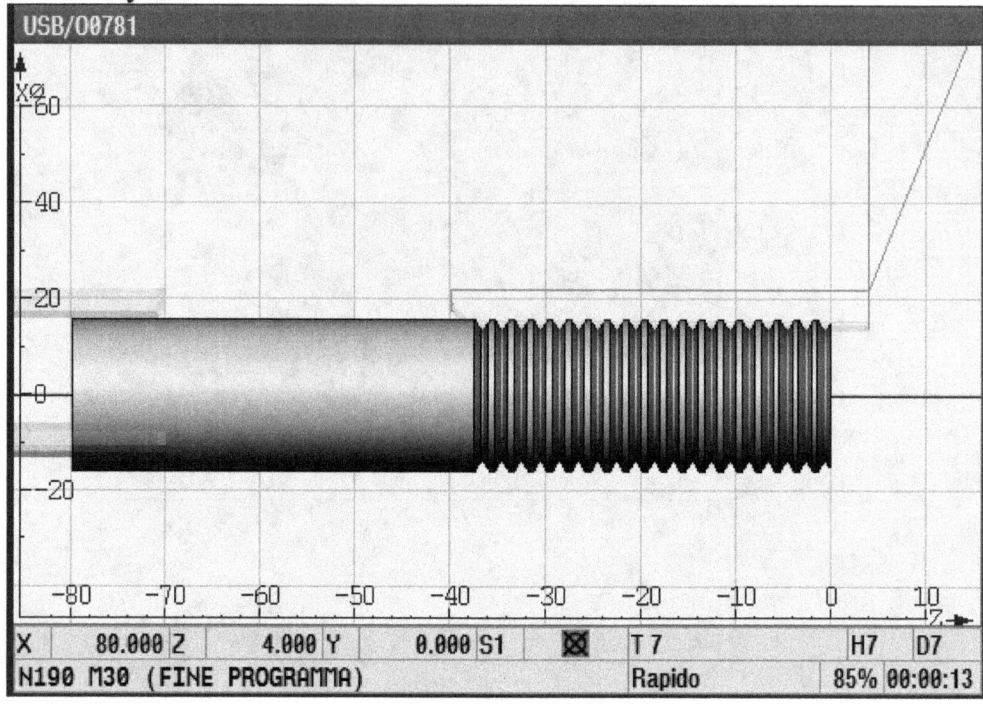

Fig. 13. G78: programming example

```
%
O0781
(G78 CYLINDRICAL THREAD)
G290 ;SIEMENS LANGUAGE ACTIVATION
WORKPIECE(,,,"CYLINDER",192,0,-80,-70,16)
G291 ; FANUC LANGUAGE ACTIVATION
G18 (X-Z PLANE)
G90 (ABSOLUTE PROGRAMMING)

T0707 (EXTERNAL THREADING TOOL)
G97 S1000 M3 (SPINDLE CONSTANT REVOLUTION)
G95 (WORKING FEED RATE IN MM/REV)

G0 X22 Z4 (CYCLE START POINT)

G78 X15.4 Z-40 R0 F2 Q0 (0.3 RADIAL DEPTH OF CUT)
X14.9 (0.25 RADIAL DEPTH OF CUT)
X14.5 (0.20 RADIAL DEPTH OF CUT)
X14.2 (0.15 RADIAL DEPTH OF CUT)
X13.96 (0.12 RADIAL DEPTH OF CUT)
```

```
X13.86 (0.05 RADIAL DEPTH OF CUT)
X13.80 (0.03 RADIAL DEPTH OF CUT)
G0 X80 (CYCLE CANCELATION)
M5
M30
%
```

4.5.2 Tapered thread

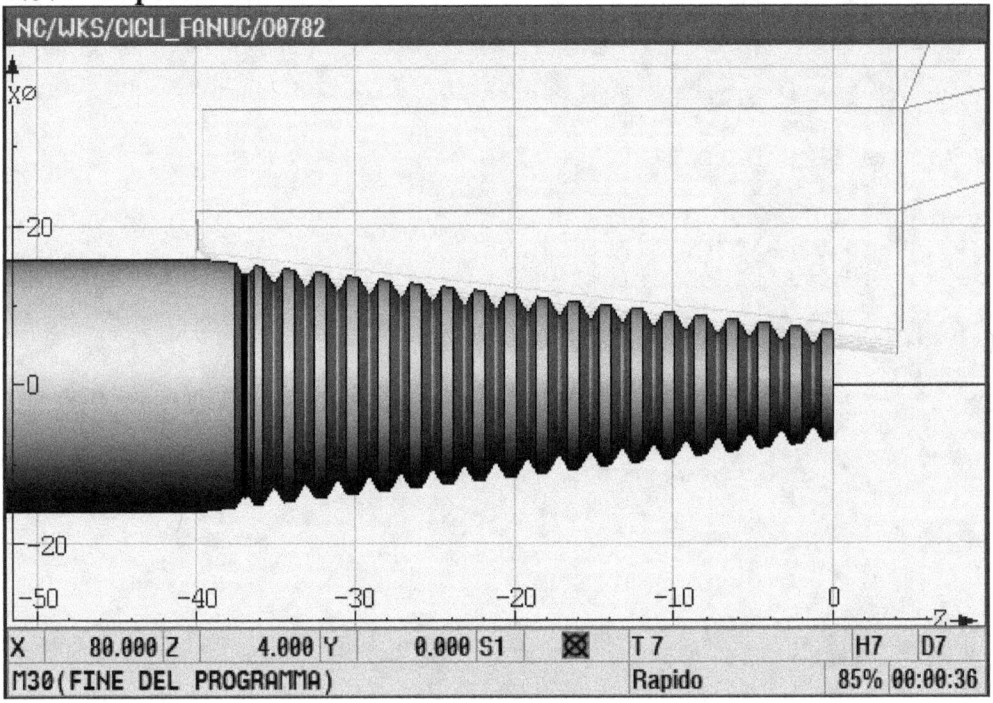

Fig. 14. G78: programming example

```
%
O0782
(G78 TAPERED THREAD)
G290 ;SIEMENS LANGUAGE ACTIVATION
WORKPIECE(,,,"CYLINDER",0,0,-110,-100,16)
G291 ;FANUC LANGUAGE ACTIVATION
G18 (X-Z PLANE)
G90 (ABSOLUTE PROGRAMMING)

T0101 (EXTERNAL TURNING TOOL)
G97 S1000 M4 (SPINDLE CONSTANT REVOLUTION)
G0 X34 Z4 (CYCLE START POINT)
G95 (WORKING FEED RATE IN MM/REV)
G77 X16 Z-40 R-5 F0.15
```

```
G0 X80 Z100 (CYCLE CANCELATION)

M5 (STOP THE SPINDLE TO REVERSE THE ROTATION)

T0707 (EXTERNAL THREADING TOOL)
G97 S1000 M3 (SPINDLE CONSTANT REVOLUTION)
G95 (WORKING FEED RATE IN MM/REV)

G0 X22 Z4 (CYCLE START POINT)
G78 X15.4 Z-40 R-5 F2 Q0 (0.3 RADIAL DEPTH OF CUT)

X14.9 (0.25 RADIAL DEPTH OF CUT)
X14.5 (0.20 RADIAL DEPTH OF CUT)
X14.2 (0.15 RADIAL DEPTH OF CUT)
X13.96 (0.12 RADIAL DEPTH OF CUT)
X13.86 (0.05 RADIAL DEPTH OF CUT)
X13.80 (0.03 RADIAL DEPTH OF CUT)
G0 X80 (CYCLE CANCELATION)

M5 (STOP SPINDLE ROTATION)
M30 (END OF PROGRAM)
%
```

5. G79: single cut facing cycle
(G94-A, G24-C)

5.1 Description

This cycle performs a single turning cut parallel to the X-axis, starting from the point where the tool is located before the cycle.
The cycle carries out four movements.
1. Rapid movement along the Z-axis, from the point programmed before the cycle to the Z-coordinate programmed in block G79.
2. Working movement along the X-axis, from the current point to the X-coordinate programmed in block G79.
3. Working movement from the end of the cut to the starting Z-coordinate programmed before the cycle.
4. Rapid return movement to the starting X-coordinate programmed before the cycle.

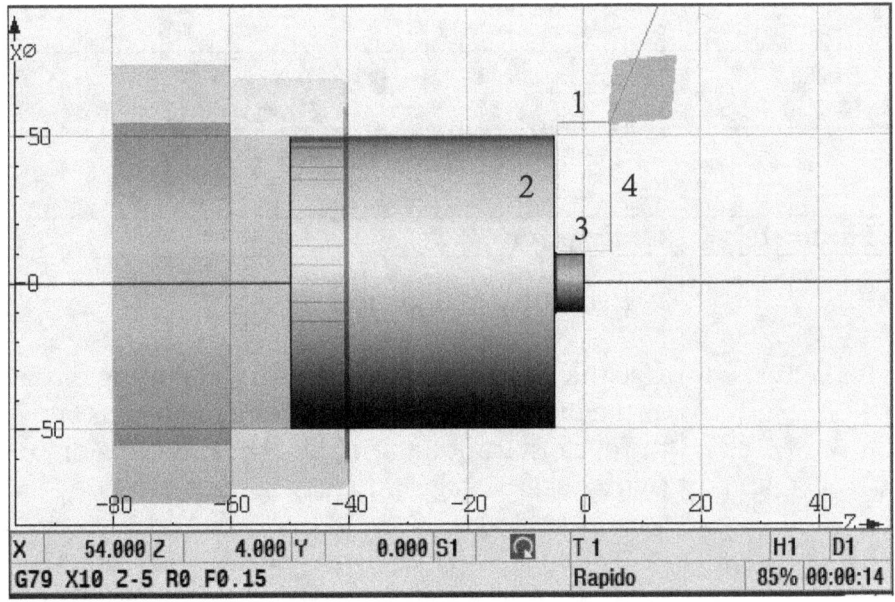

Fig. 15. G79: cycle movements

5.2 Cycle cancelation functions

Program a G-code of group 01 to cancel this cycle. If the machine uses the "A-code" system, the G-code must be different from G90, G92 or G94.

5.3 Cycle parameters

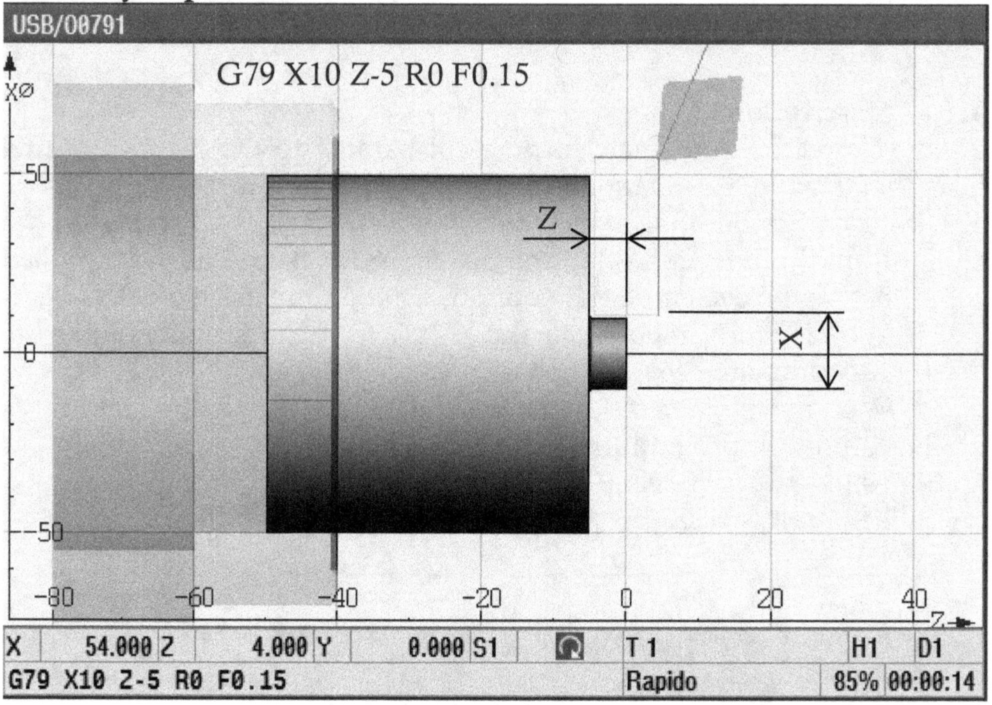

Fig. 16. G79: cycle parameters

Parameter	Description
X	Arrival cutting diameter.
(U)	Use the letter "U" to set the arrival cutting diameter as the diametrical distance along the X-axis between the cycle start point and the arrival cutting diameter (with negative sign for external cuts).

Z	Z position of the facing cut referring to the workpiece zero point.
(W)	Use the letter "W" to set the cutting position as the distance along the Z-axis from the point programmed before the cycle to the starting point of the cut.
Notes	In case of taper facing, this coordinate is related to the position of the arrival cutting point (the starting point will be calculated by the cycle using the parameter "R").

To be continued

R	Taper value. It is equal to the starting position in Z minus the arrival position in Z of the cut. The starting and arrival positions of the cut might not correspond to the starting/arrival positions of the taper, but to the positions of the cut to be performed, calculated on the basis of the tool position in X at the beginning and at the end of the cycle. In the following picture, the R value is negative. When the value is zero or when the parameter is not programmed, facing is performed parallel to the X-axis. 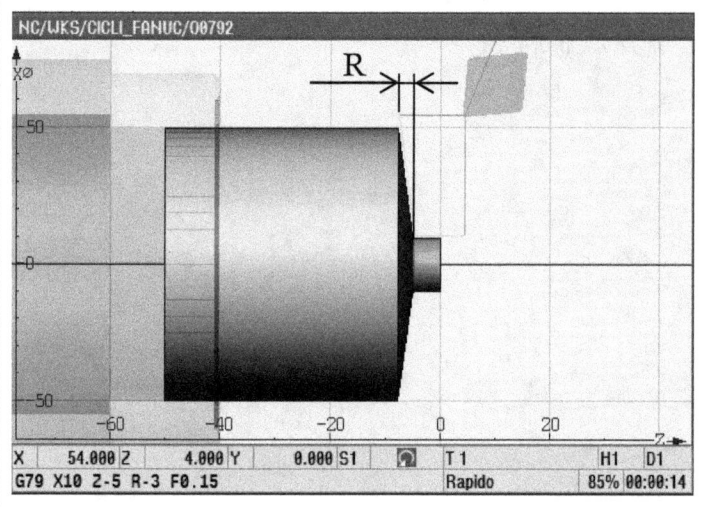
F	Working feed rate.

5.4 Programming examples

5.4.1 Straight face

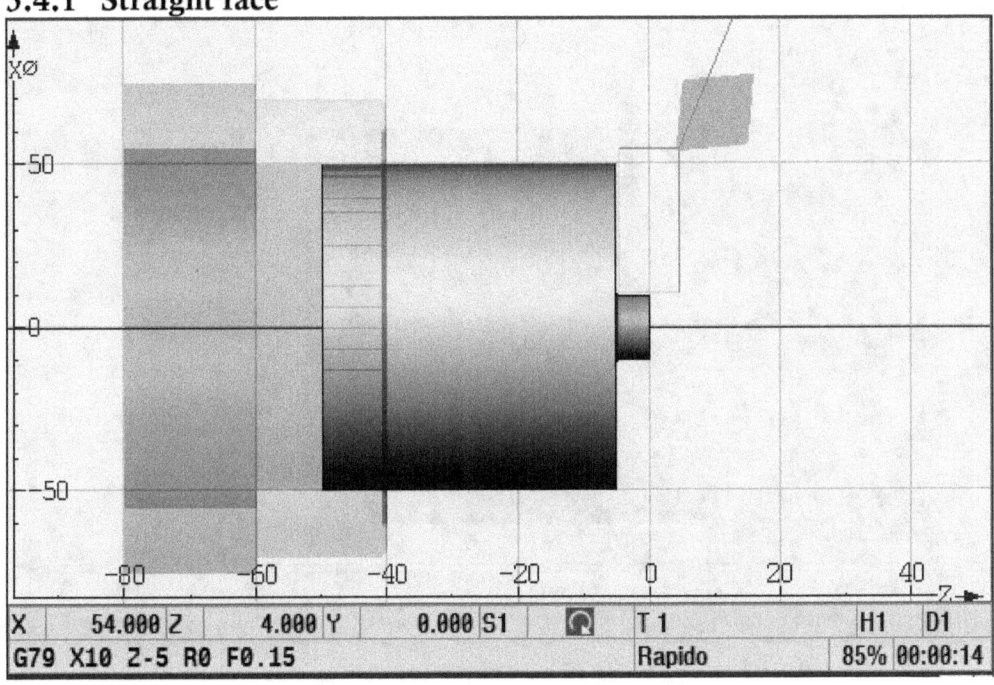

Fig. 17. G79: programming example

```
%
O0791
(G79 FACING CUT)
G290 ;SIEMENS LANGUAGE ACTIVATION
WORKPIECE(,,,"CYLINDER",192,0,-50,-40,50)
G291 ;FANUC LANGUAGE ACTIVATION
G18 (X-Z PLANE)
G90 (ABOSULTE PROGRAMMING)

T0101 (EXTERNAL TURNING TOOL)
G92 S3000 (SPINDLE RPM MAX LIMIT)
G96 S100 M4 (CONSTANT CUTTING SPEED)
G95 (WORKING FEED RATE IN MM/REV)

G0 X54 Z4 (CYCLE START POINT)
G79 X10 Z-5 R0 F0.15
G0 X80 (CYCLE CANCELATION)
M5 (STOP SPINDLE ROTATION)
M30 (END OF PROGRAM)
%
```

5.4.2 Sloped shoulder

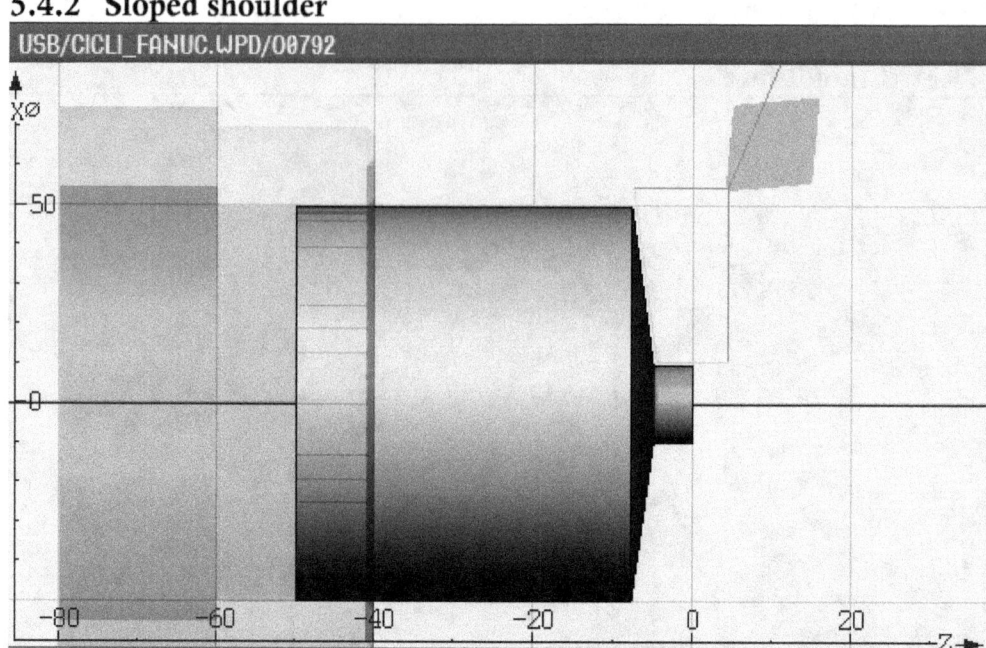

Fig. 18. G79: programming example

```
%
O0792
(G79 SLOPED FACING CUT)
G290 ;SIEMENS LANGUAGE ACTIVATION
WORKPIECE(,,,"CYLINDER",192,0,-50,-40,50)
G291 ; FANUC LANGUAGE ACTIVATION
G18 G90 (X-Z PLANE, ABSOLUTE PROGRAMMING)

T0101 (EXTERNAL TURNING TOOL)
G92 S3000 (SPINDLE RPM MAX LIMIT)
G96 S100 M4 (CONSTANT CUTTING SPEED)
G95 (WORKING FEED RATE IN MM/REV)

G0 X54 Z4 (CYCLE START POINT)
G79 X10 Z-5 R-3 F0.15
G0 X80 (CYCLE CANCELATION)
M5 (STOP SPINDLE ROTATION)
M30 (END OF PROGRAM)
%
```

In this example, X54 and Z-8 are the coordinates of the starting point of the cut.

6. G70: finishing cycle or profile repetition
(G70-A, G72-C)

6.1 Description

This cycle repeats part of a program comprised between two block numbers.

G70 is also called **finishing cycle** because it is often programmed after a roughing cycle for the contour finishing.

At the end of the profile, the cycle takes the tool back to the point programmed before the cycle; this point therefore has to be in a position that does not interfere with the material during the last retraction. Therefore, a point outside the workpiece is recommended.

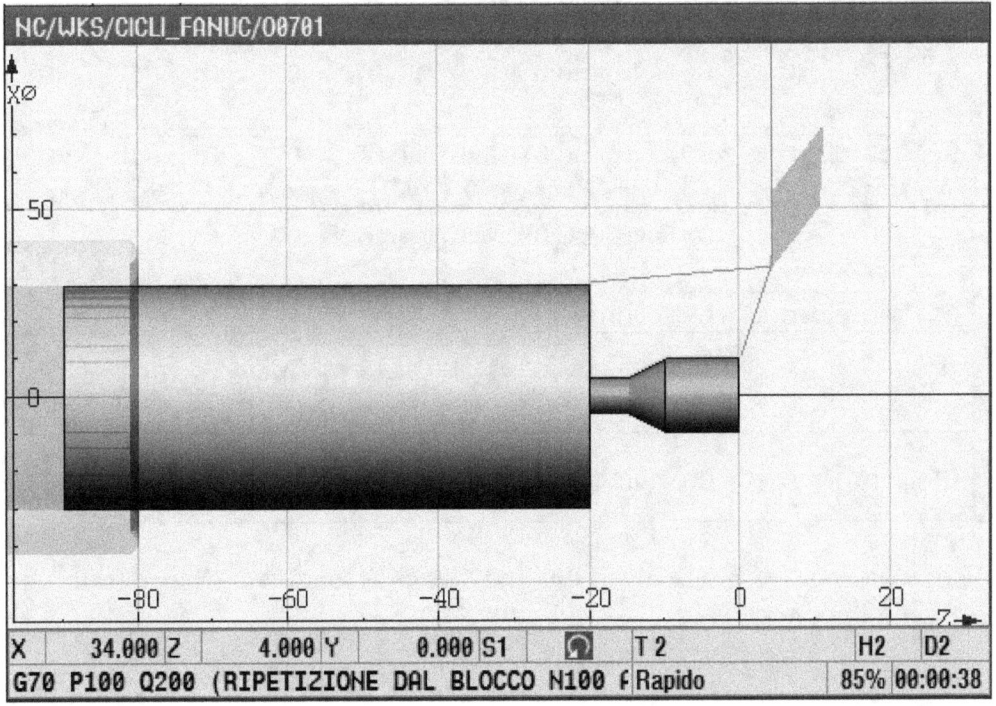

Fig. 19. G70: cycle movements

6.2 Cycle cancelation functions

This cycle is self-canceling: no further functions are necessary to deactivate it.

6.3 Cycle parameters

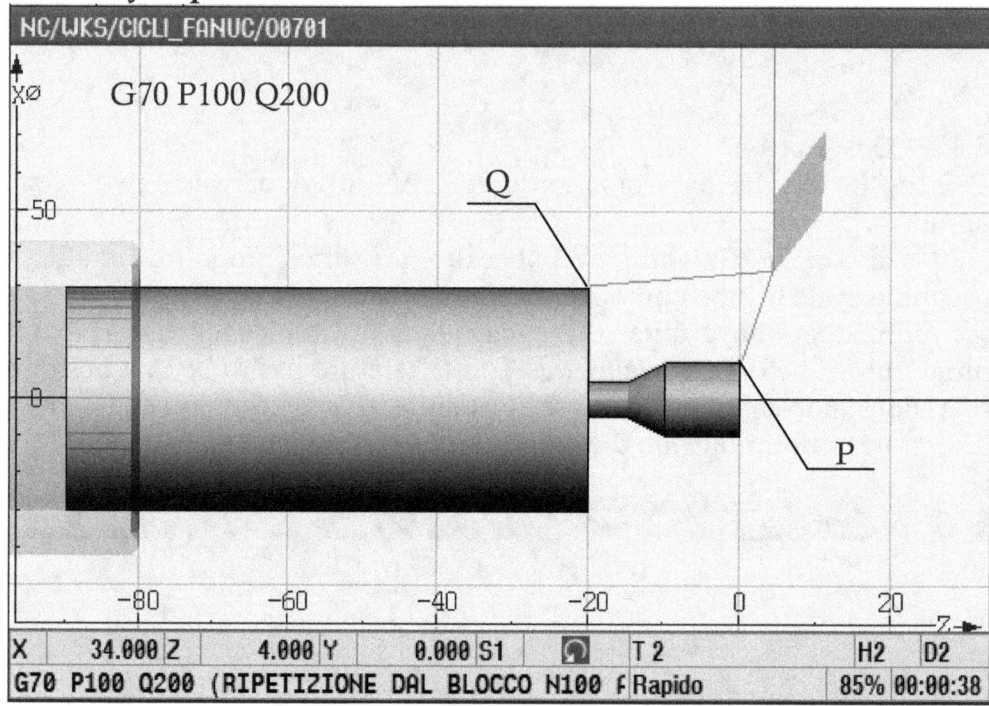

Fig. 20 G70: cycle parameters

Parameter	Description
P	First block number of the finishing profile.
Q	Last block number of the finishing profile.

The working feed rate used by the cycle is enabled or programmed between the parameters "P" and "Q".

6.4 Programming examples

6.4.1 Finishing

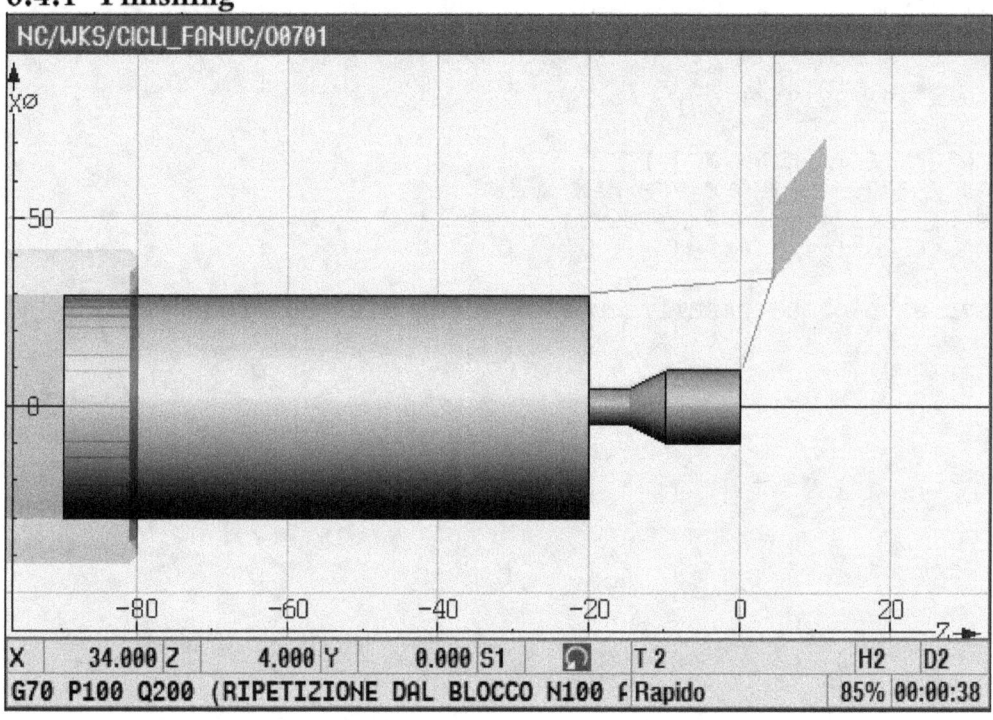

Fig. 21. G70: programming example

```
%
O0701
(G70 FINISHING CYCLE)
G290 ;SIEMENS LANGUAGE ACTIVATION
WORKPIECE(,,,"CYLINDER",192,0,-90,-80,30)
G291 ;FANUC LANGUAGE ACTIVATION
G18 (X-Z PLANE)
G90 (ABSOLUTE PROGRAMMING)

(ROUGHING)
T0202 (EXTERNAL TURNING TOOL)
G92 S3000 (SPINDLE RPM MAX LIMIT)
G96 S100 M4 (CONSTANT CUTTING SPEED)
G95 (WORKING FEED RATE IN MM/REV)

G0 X30 Z4 (POINT BEFORE THE CYCLE)
G71 U2 R1
G71 P100 Q200 U1 W0.1 F0.15

(PROFILE PROGRAMMING)
```

```
N100 G0 X10 Z0 (FIRST CONTOUR BLOCK)
G1 Z-10 F0.08
G1 Z-15 X5
G1 Z-20
N200 G1 X30 (LAST CONTOUR BLOCK)

G0 X150 (RETRACTION)

T0202 (FINISHING TOOL)
G92 S3000 (SPINDLE RPM MAX LIMIT)
G96 S120 M4 (CONSTANT CUTTING SPEED)
G0 Z4 X34

G70 P100 Q200 (REPETITION FROM BLOCK N100 TO BLOCK N200)

G0 X200
G0 Z200
M5
M30
%
```

6.4.2 Groove repetition

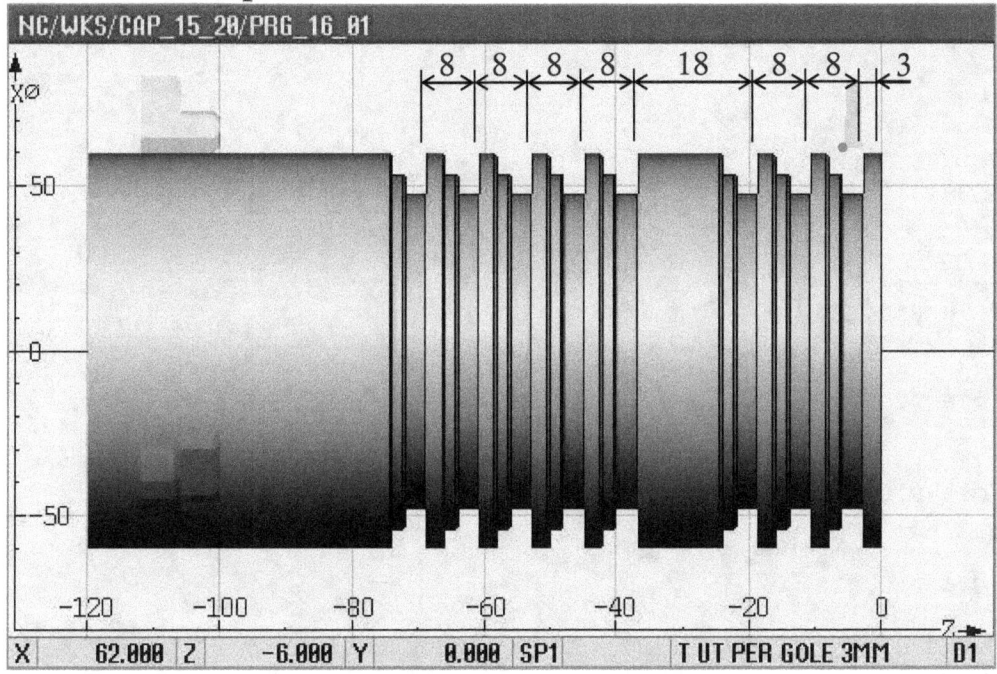

Fig. 22. G70: programming example

```
%
O0702
(G70 GROOVE REPETITION)
G290 ;SIEMENS LANGUAGE ACTIVATION
WORKPIECE(,,,"CYLINDER",0,0,-120,-100,60)
G291 ;FANUC LANGUAGE ACTIVATION
G18 (X-Z PLANE)
G90 (ABSOLUTE PROGRAMMING)

(GROOVE 1)
T1010 (GROOVING TOOL, INSERT WIDTH 3MM, THE TOOL GEOMETRY IS
ON THE LEFT SIDE OF THE INSERT)
G97 S1150 M4

(GROOVE ROUGHING)
G0 X62 Z-6

N10 G1 X48.1 F0.12
G1 X62 F1
G91 G1 Z-2
G90 G1 X54.1 F0.12
G1 X62 F1
```

```
(CHAMFERS AND PROFILE FINISHING)
G1 G91 Z-0.5
G90 G1 X60 F0.12
G91 G1 Z0.5 ,A-45
G90 G1 X54
G91 G1 Z1.5
G1 Z0.5 ,A-45
G90 G1 X48
N20 G1 X62 F1

G0 Z-14 (GROOVE 2 START POINT)
G70 P10 Q20

G0 Z-22 (GROOVE 3 START POINT)
G70 P10 Q20

G0 Z-40 (GROOVE 4 START POINT)
G70 P10 Q20

G0 Z-48 (GROOVE 5 START POINT)
G70 P10 Q20

G0 Z-56 (GROOVE 6 START POINT)
G70 P10 Q20

G0 Z-64 (GROOVE 7 START POINT)
G70 P10 Q20

G0 Z-72 (GROOVE 8 START POINT)
G70 P10 Q20

G0 X200
G0 Z200

M5
M30
%
```

7. G71: roughing cycle along the Z-axis
(G71-A, G73-C)

7.1 Description

This cycle entirely removes excess stock from the profile with cuts parallel to the Z-axis, if necessary leaving an allowance for the contour finishing. The cycle allows to set the cut depth with a radial value, the retraction distance at the end of the cut, the first and the last block of the profile to be roughed, the allowance to be left for finishing and the specific working feed rate used during the roughing passes.

It is important to underline that by programming the first block of the profile to be roughed, two different types of cycle application are determined. These two types are called "type 1" and "type 2" in Fanuc manuals.

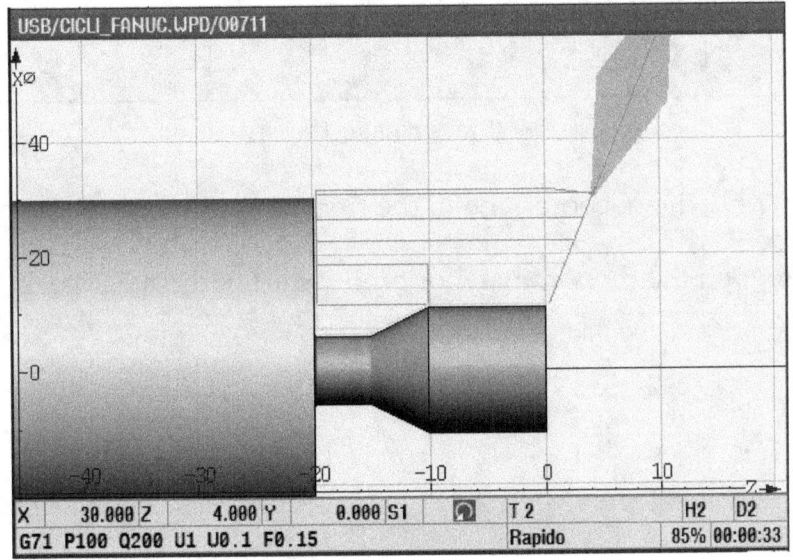

Fig. 23. G71: cycle movements

7.1.1 Roughing cycle type 1
Roughing cycle type 1 is characterized by the following features:
- It does not perform the contouring of shaded profile parts. Shaded parts are decreasing parts called "pockets" (in Fanuc manuals, this kind of profile is called "non monotonous").
- The cutter is retracted at a 45° angle at the end of the cut.
- The cutter approaches the stock material starting from the Z-coordinate programmed before the cycle.

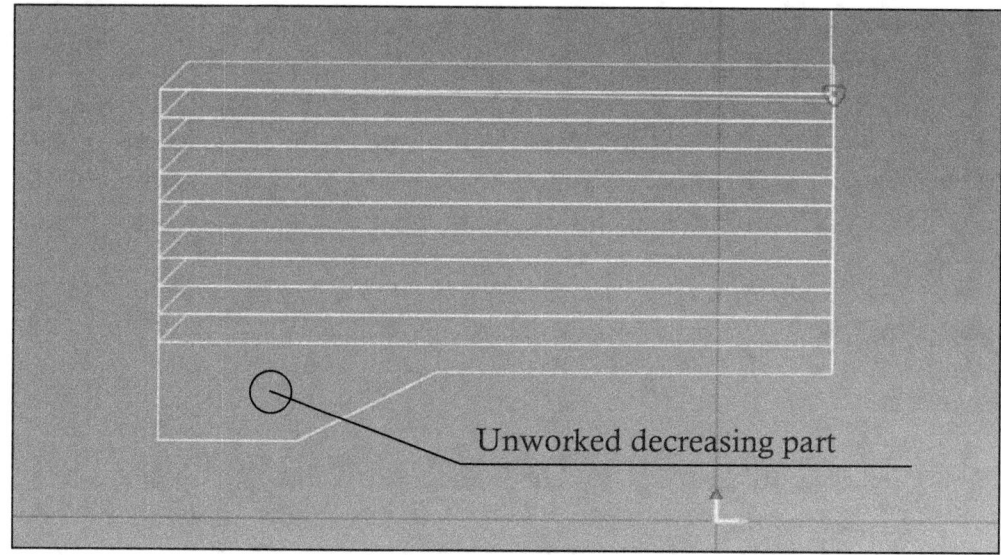

Fig. 24. G71: roughing cycle type 1

Program only the X-coordinate in the first contour block to set roughing cycle type 1.
Then program the Z-coordinate for profile start in the next block.

```
N10 G1 X10 (CYCLE TYPE 1)
G1 Z0
G1 Z-10
G1 Z-15 X5
G1 Z-20
N20 G1 X30
```

7.1.2 Roughing cycle type 2

Roughing cycle type 2 is characterized by the following features:
- It performs the contouring of shaded parts of the profile.
- The cutter follows the profile at the end of the cut to remove the material left by the tool shape.
- The cutter approaches the stock material and closes the gap between the Z-coordinate programmed before the cycle and point Z for profile start.

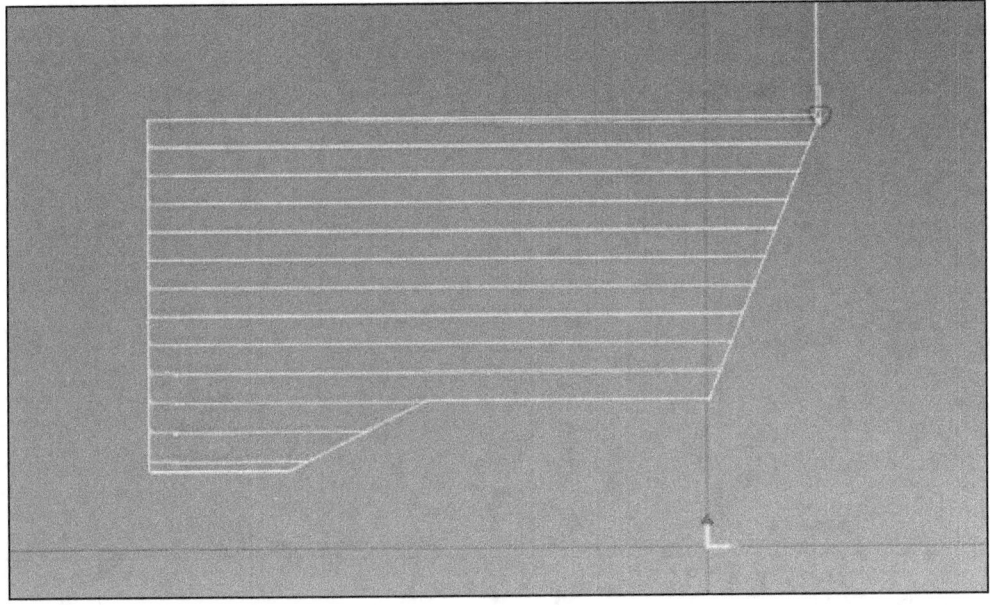

Fig. 25. G71: roughing cycle type 2

Program both coordinates (X and Z) in the first contour block to set roughing cycle type 2.

```
N10 G1 X10 Z0  (CYCLE TYPE 2)
G1  Z-10
G1  Z-15 X5
G1  Z-20
N20 G1 X30
```

7.1.3 Setting of the approaching feed rate
In both type 1 and type 2 roughing cycles:
- set G0 in the first block to set a rapid approach before cutting
- set G1 in the first block to approach with a working feedrate.

```
N10 G0 X10 (G1 X10)
G1 Z0
G1 Z-10
G1 Z-15 X5
G1 Z-20
N20 G1 X30
```

7.2   Cycle cancelation functions
This cycle is self-canceling: no further functions are necessary to deactivate it.

7.3   Cycle-related NC parameters
The pre-finishing operation consists of a continuous cut along the finishing contour after roughing.
It may be useful in cycle type 1 to level all shoulders.
It may be a waste of time in cycle type 2 because the shoulders are already flattened.
The pre-finishing cut can be cancelled changing the following machine parameter.

Parameter	Description
N. 5105.1	Bit RF1 = 0 The pre-finishing cut is performed = 1 The pre-finishing cut is not performed

7.4 Cycle parameters

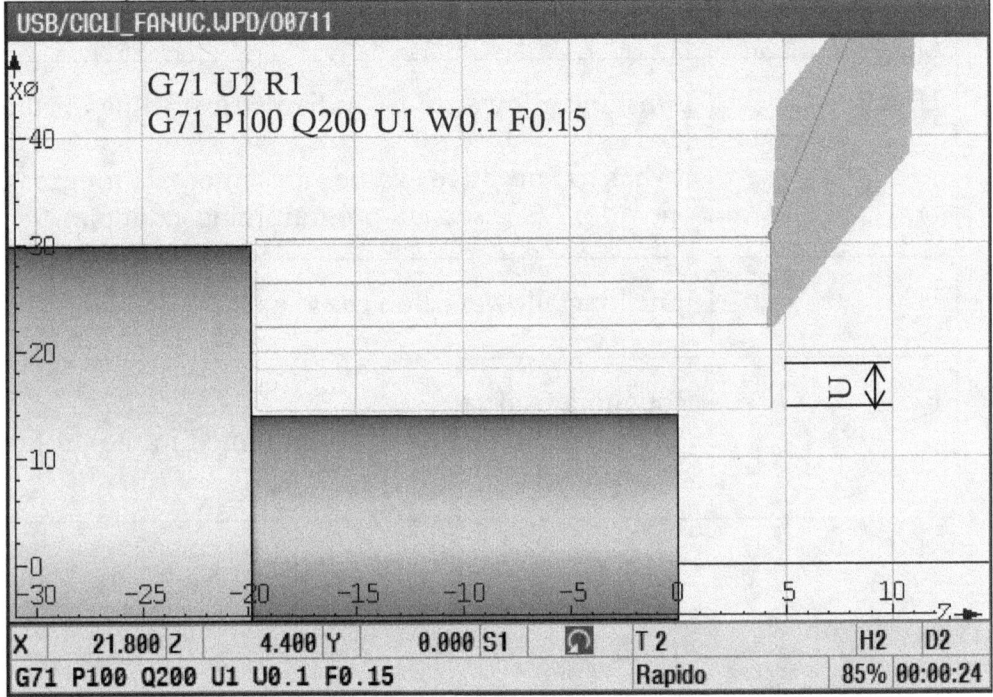

Fig. 26. G71: cycle parameters

First block

Parameter	Description
U	Radial depth of cut. It has a negative value for internal roughing operations (see second programming example).
R	Retraction distance along the X- and Z-axis at the end of the cut.

Second block

Parameter	Description
P	First block number of the profile.

Q	Last block number of the profile.
U	Finishing allowance with diametrical value on the X-axis. It has a negative value for internal roughing operations (see second programming example).
W	Finishing allowance on the Z-axis.
F	Roughing feed rate. Any working feed rates programmed in the profile are ignored.

7.5 Programming examples

7.5.1 External rough turning

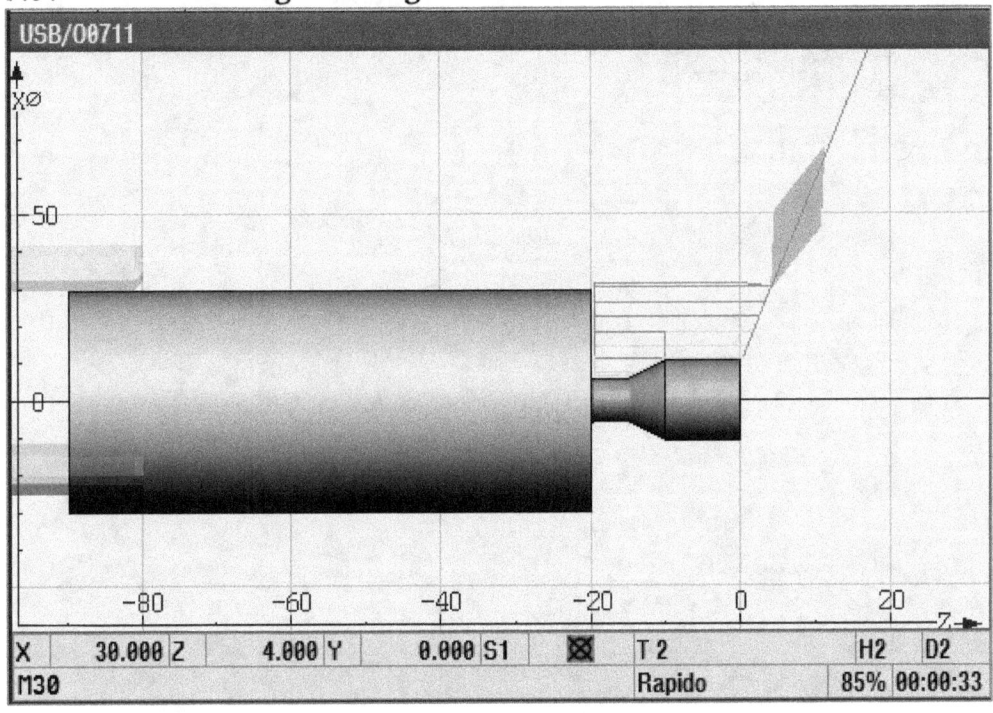

Fig. 27. G71: programming example

```
%
O0711
(G71 ROUGHING CYCLE ALONG Z-AXIS)
G290 ;SIEMENS LANGUAGE ACTIVATION
WORKPIECE(,,,"CYLINDER",192,0,-90,-80,30)
G291 ;FANUC LANGUAGE ACTIVATION
G18 (X-Z PLANE)
G90 (ABSOLUTE PROGRAMMING)

(ROUGHING)
T0202 (EXTERNAL TURNING TOOL)
G92 S3000 (SPINDLE RPM MAX LIMIT)
G96 S100 M4 (CONSTANT CUTTING SPEED)
G0 X30 Z4 (POINT BEFORE THE CYCLE)
G95 (WORKING FEED RATE IN MM/REV)
G71 U2 R1
G71 P100 Q200 U1 W0.1 F0.15

(PROFILE PROGRAMMING)
N100 G0 X10 Z0 (FIRST BLOCK OF THE PROFILE)
```

```
G1 Z-10
G1 Z-15 X5
G1 Z-20
N200 G1 X30 (LAST BLOCK OF THE PROFILE)

G0 X200 (RETRACTION)
G0 Z200
M5
M30
%
```

7.5.2 Internal rough turning

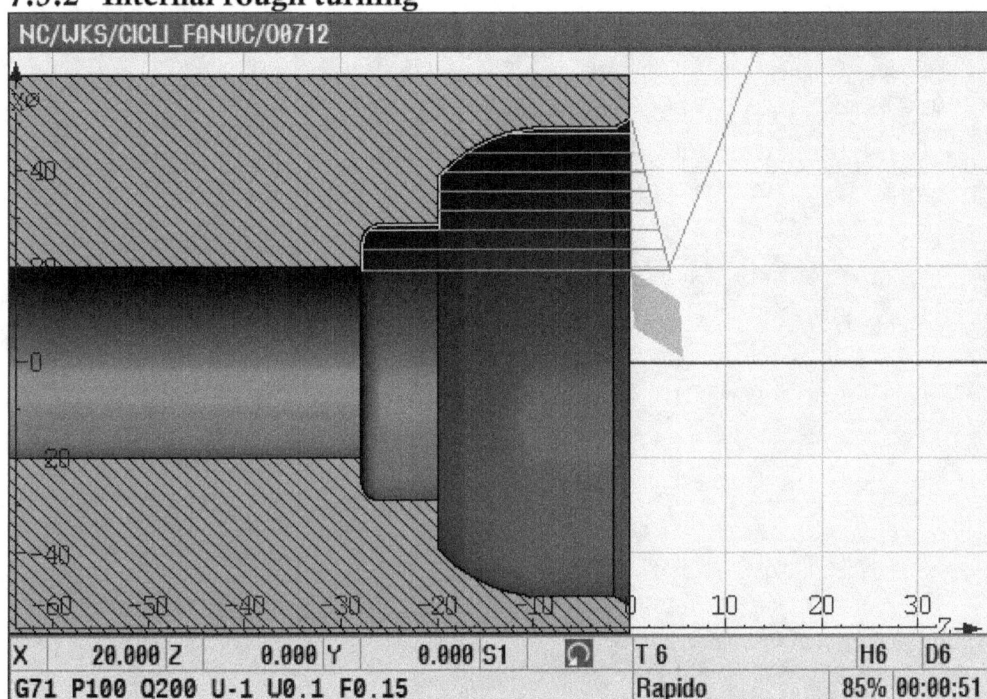

Fig. 28. G71: programming example

```
%
O0712
(G71 INTERNAL ROUGH TURNING ALONG Z-AXIS)
G290 ;SIEMENS LANGUAGE ACTIVATION
WORKPIECE(,,,"PIPE",192,0,-90,-80,60,20)
G291 ;FANUC LANGUAGE ACTIVATION
G18 (X-Z PLANE)
G90 (ABSOLUTE PROGRAMMING)

(ROUGHING)
T0606 (INTERNAL TURNING TOOL)
G92 S3000 (SPINDLE RPM MAX LIMIT)
G96 S80 M4 (CONSTANT CUTTING SPEED)
G95 (WORKING FEED RATE IN MM/REV)

G0 X20 Z4 (POINT BEFORE THE CYCLE)
G71 U-2 R1
G71 P100 Q200 U-1 W0.1 F0.15

(PROFILE PROGRAMMING)
N100 G0 X52 Z0 (FIRST BLOCK OF THE PROFILE)
G1 X50 ,A210 F0.1
```

```
G1 Z-10
G3 X40 Z-20 R15
G1 X30
G1 Z-28 ,R2
N200 G1 X20 (LAST BLOCK OF THE PROFILE)

G0 Z200 (RETRACTION)
G0 X200
M5
M30
%
```

8. G72: roughing cycle along the X-axis
(G72-A, G74-C)

8.1 Description

This cycle entirely removes excess stock from the profile with cuts parallel to the X-axis, if necessary leaving an allowance for the contour finishing.

The cycle allows to set the cut width, the retraction distance at the end of the cut, the first and the last block of the profile to be roughed, the allowance to be left for finishing and the specific working feed rate used during the roughing passes.

As already seen for cycle G71, also G72, by programming the first block of the profile to be roughed, two different types of cycle application are determined. These two types are called "type 1" and "type 2".

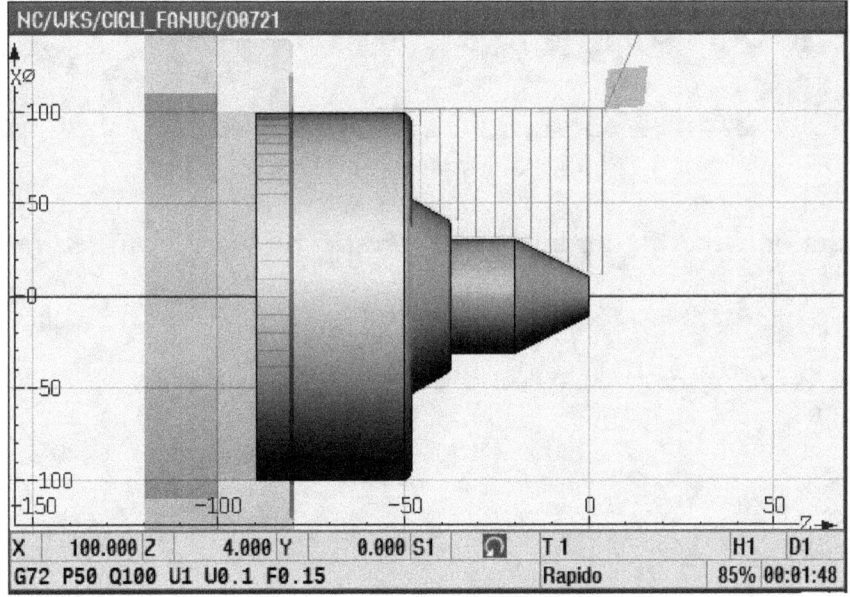

Fig. 29. G72: cycle movements

8.1.1 Roughing cycle type 1

Roughing cycle type 1 is characterized by the following features:
- It does not perform the contouring of shaded profile parts. Shaded parts are decreasing parts called "pockets" (in Fanuc manuals, this kind of profile is called "non monotonous").
- The cutter retracts at a 45° angle at the end of the cut.
- The cutter approaches the stock material starting from the Z-coordinate programmed before the cycle.

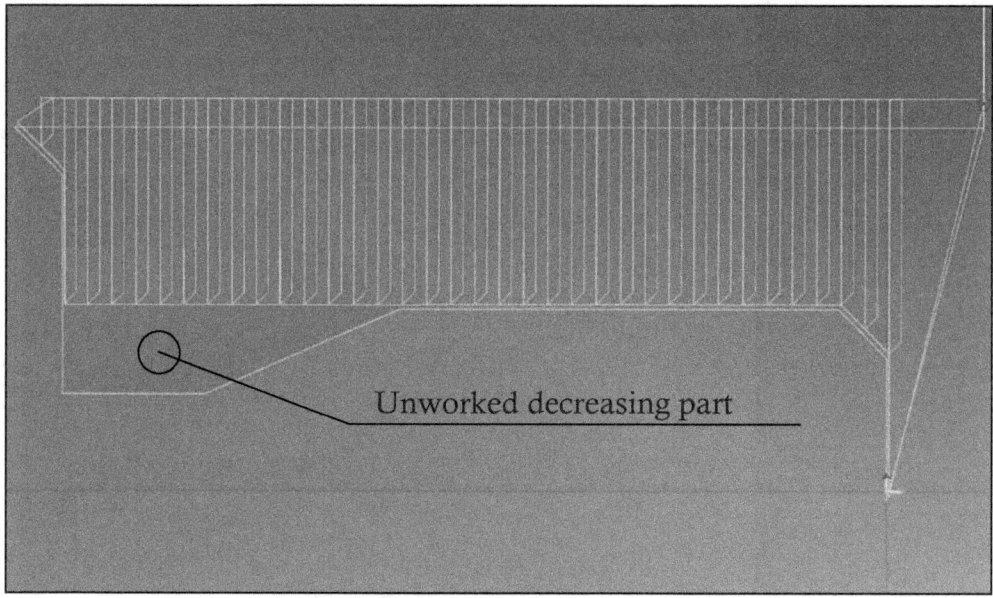

Fig. 30. G72: roughing cycle type 1

Program only the X-coordinate in the first contour block to set roughing cycle type 1.
Then, program the Z-coordinate for profile start in the next block.

```
N50 G1 X100 (CYCLE TYPE 1)
G1 Z-58
G1 Z-48 ,R4
G1 X52
G1 X40 ,A330 ,R2
G1 X30
G1 Z-20
N100 G1 X10 Z0
```

8.1.2 Roughing cycle type 2

Roughing cycle type 2 is characterized by the following features:
- It performs the contouring of shaded parts of the profile.
- The cutter follows the profile at the end of the cut to remove the material left by the tool shape.
- The cutter approaches the stock material and closes the gap between the Z-coordinate programmed before the cycle and point Z for profile start.

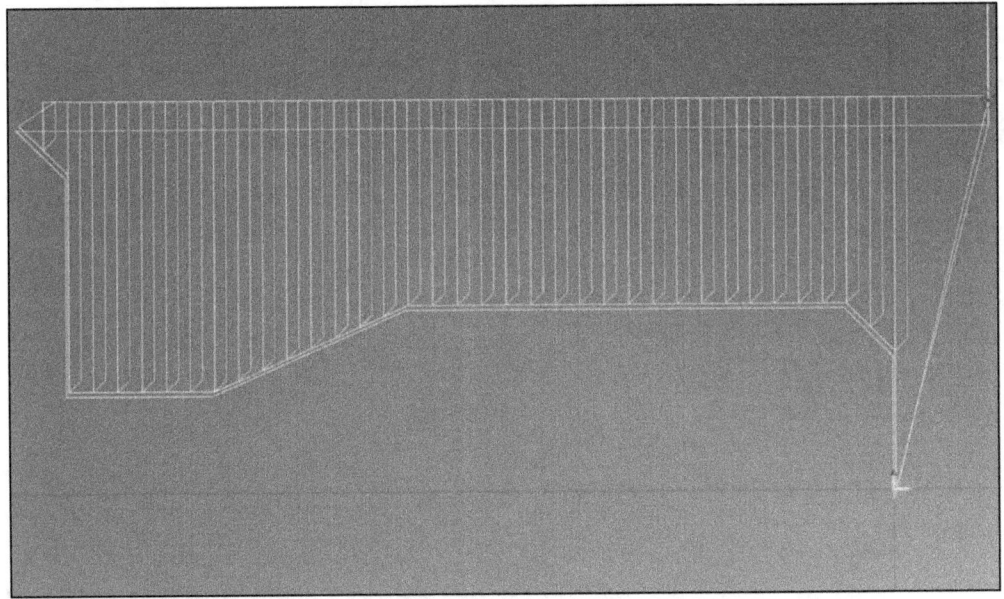

Fig. 31. G72: roughing cycle type 2

Program both coordinates (X and Z) in the first contour block to set roughing cycle type 2.

```
N50 G1 X100 Z-58  (CYCLE TYPE 2)
G1  Z-48 ,R4
G1  X52
G1  X40 ,A330 ,R2
G1  X30
G1  Z-20
N100 G1 X10 Z0
```

8.1.3 Setting of the approaching feed rate

In both type 1 and type 2 roughing cycles:
- set G0 in the first block to set a rapid approach before cutting
- set G1 in the first block to approach with a working feedrate.

```
N50 G0 X100   (G1 X100)
G1 Z-58
G1 Z-48 ,R4
G1 X52
G1 X40 ,A330 ,R2
G1 X30
G1 Z-20
N100 G1 X10 Z0
```

8.2 Cycle cancelation functions

This cycle is self-canceling: no further functions are necessary to deactivate it.

8.3 Cycle-related NC parameters

The pre-finishing operation consists of a continuous cut along the finishing contour after roughing. The pre-finishing cut can be cancelled changing the following machine parameter.

Parameter	Description
N. 5105.1	Bit RF1 = 0 The pre-finishing cut is performed = 1 The pre-finishing cut is not performed

8.4 Cycle parameters

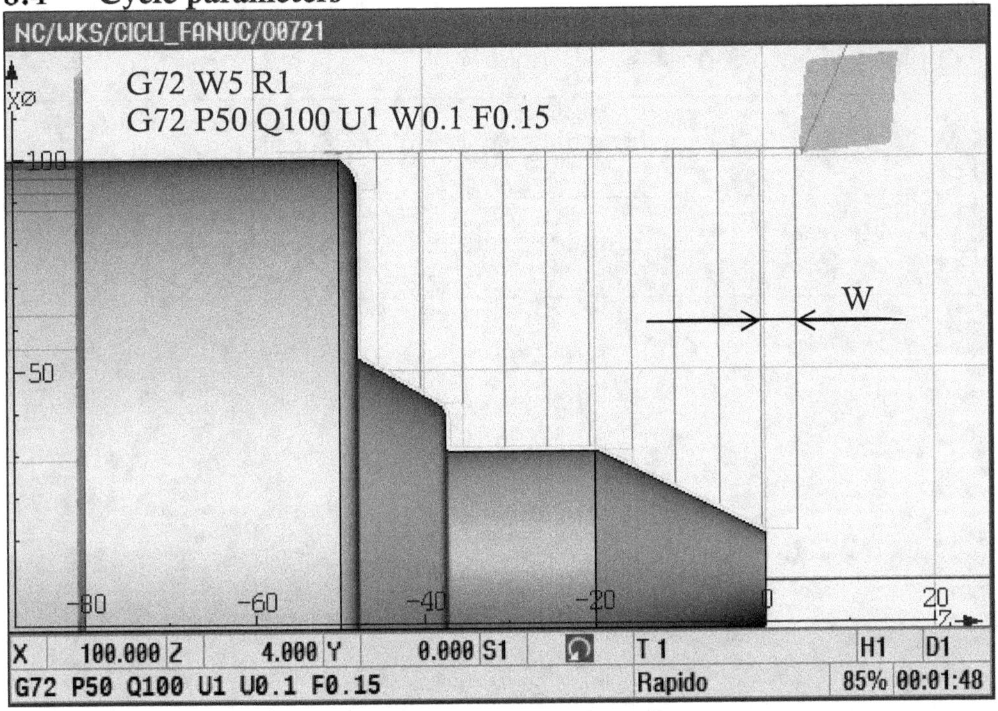

Fig. 32. G72: cycle parameters

First block

Parameter	Description
W	Cut width.
R	Retraction distance along the X- and Z-axis at the end of the cut.

Second block

Parameter	Description
P	First block number of the finishing profile.
Q	Last block number of the finishing profile.

U	Finishing allowance with diametrical value on the X-axis.
W	Finishing allowance on the Z-axis.
F	Roughing feed rate.

8.5 Programming example

8.5.1 Roughing

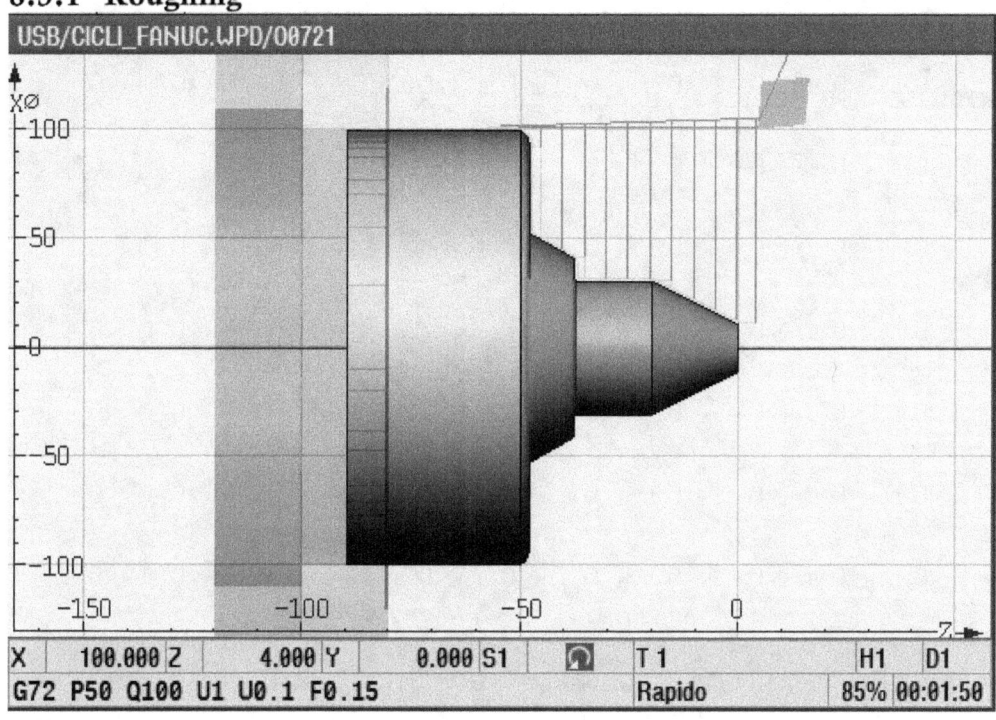

Fig. 33 .G72: programming example

```
%
O0721
(G72 ROUGHING CYCLE ALONG X-AXIS)
G290 ;SIEMENS LANGUAGE ACTIVATION
WORKPIECE(,,,"CYLINDER",192,0,-90,-80,100)
G291 ;FANUC LANGUAGE ACTIVATION
G18 (X-Z PLANE)
G90 (ABSOLUTE PROGRAMMING)

(ROUGHING)
T0101 (EXTERNAL TURNING TOOL)
G92 S3000 (SPINDLE RPM MAX LIMIT)
G96 S100 M4 (CONSTANT CUTTING SPEED)
G95 (WORKING FEED RATE IN MM/REV)

G0 X104 Z4 (POINT BEFORE THE CYCLE)
G72 W5 R1
G72 P50 Q100 U1 W0.1 F0.15

(PROFILE PROGRAMMING)
```

```
N50 G0 X100 Z-58 (FIRST BLOCK OF THE PROFILE)
G1 Z-48 ,R4
G1 X52
G1 X40 ,A330 ,R2
G1 X30
G1 Z-20
N100 G1 X10 Z0 (LAST BLOCK OF THE PROFILE)

G0 X200 (RETRACTION)
G0 Z200
M5
M30
%
```

9. G73: pattern repeating cycle
(G73-A, G75-C)

9.1 Description

This cycle is ideal for roughing molded work pieces. It entirely removes excess stock from the profile to be created with cuts parallel to the profile itself, if necessary, leaving an allowance for the contour finishing.

The cycle allows to set the amount of stock to be removed along the X- and Z-axis, the number of profile repetitions (which will determine the cut depth), the finishing allowance and the specific working feed rate used during the roughing passes.

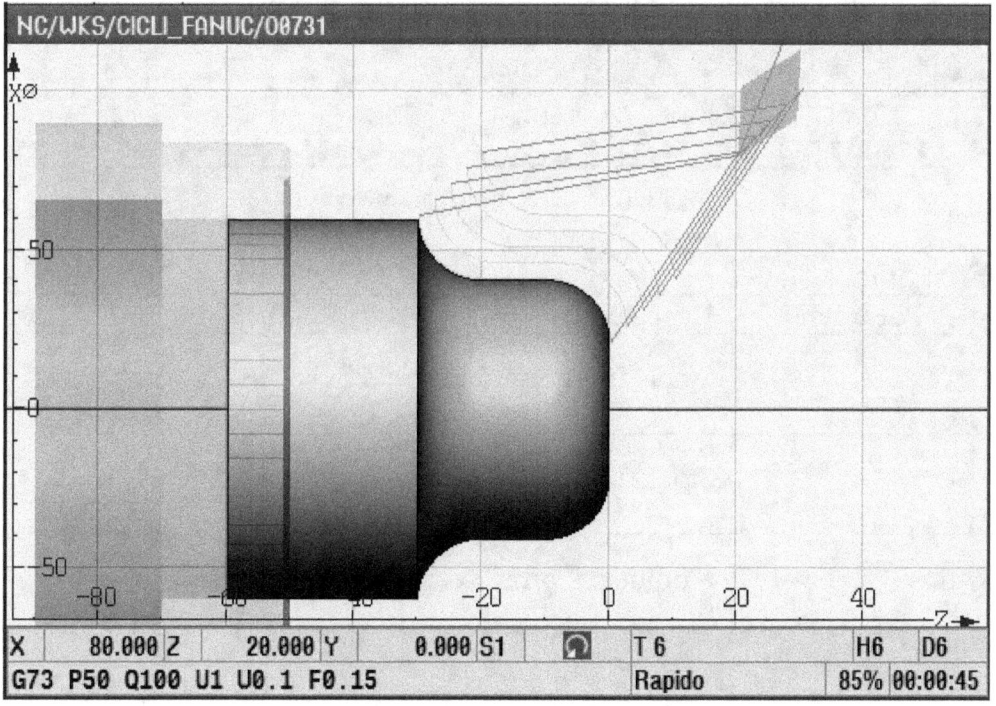

Fig. 34. G73: cycle movements

9.2 Cycle cancelation functions

This cycle is self-canceling: no further functions are necessary to deactivate it.

9.3 Cycle parameters

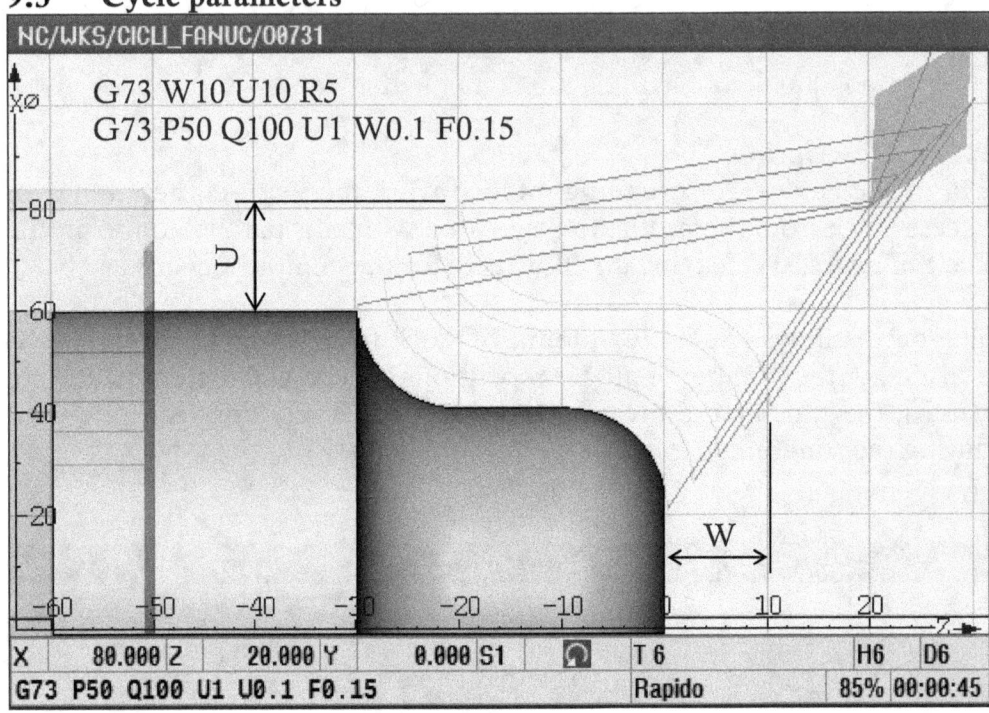

Fig. 35. G73: cycle parameters

First block

Parameter	Description
W	Stock amount along the Z-axis.
U	Stock amount along the X-axis (radial value).
R	Number of profile repetitions.

Second block

Parameter	Description
P	First block number of the finishing profile.
Q	Last block number of the finishing profile.
U	Finishing allowance with diametrical value along the X-axis.
W	Finishing allowance on the Z-axis.
F	Roughing feed rate.

9.4 Programming example

9.4.1 Roughing

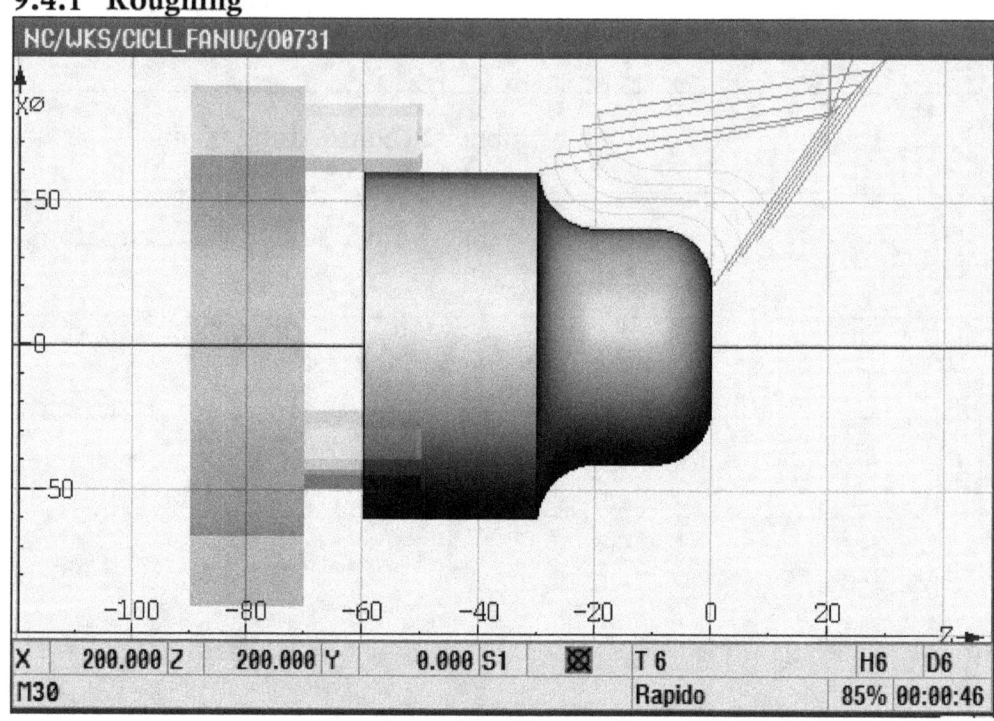

Fig. 36. G73: programming example

```
%
O0731
(G73 PATTERN REPEATING CYCLE)
G290 ;SIEMENS LANGUAGE ACTIVATION
WORKPIECE(,,,"CYLINDER",192,0,-60,-50,60)
G291 ;FANUC LANGUAGE ACTIVATION
G18 (X-Z PLANE)
G90 (ABSOLUTE PROGRAMMING)

(ROUGHING)
T0202 (EXTERNAL TURNING TOOL)
G97 S2000 M4 (SPINDLE CONSTANT REVOLUTION)
G95 (WORKING FEED RATE IN MM/REV)

G0 X80 Z20 (POINT BEFORE THE CYCLE)
G73 W10 U10 R5
G73 P50 Q100 U1 W0.1 F0.15

(PROFILE PROGRAMMING)
N50 G0 X20 Z0 (FIRST BLOCK OF THE PROFILE)
```

```
G3 X40 Z-10 R10
G1 Z-20
N100 G2 X60 Z-30 R10 (LAST BLOCK OF THE PROFILE)

G0 X200 (RETRACTION)
G0 Z200
M5
M30
%
```

72

10. G76: multi-pass threading cycle
(G76-A, G78-C)

10.1 Description

The cycle performs a complete thread in several cuts. The thread can be cylindrical or conical and must have constant lead. The distribution of the passes can take place in a radial direction or along one side of the insert. The cycle can automatically reduce the depth of cut while keeping the chip section constant. The thread exit direction of the tool at the end of the thread can be set by using a parameter.

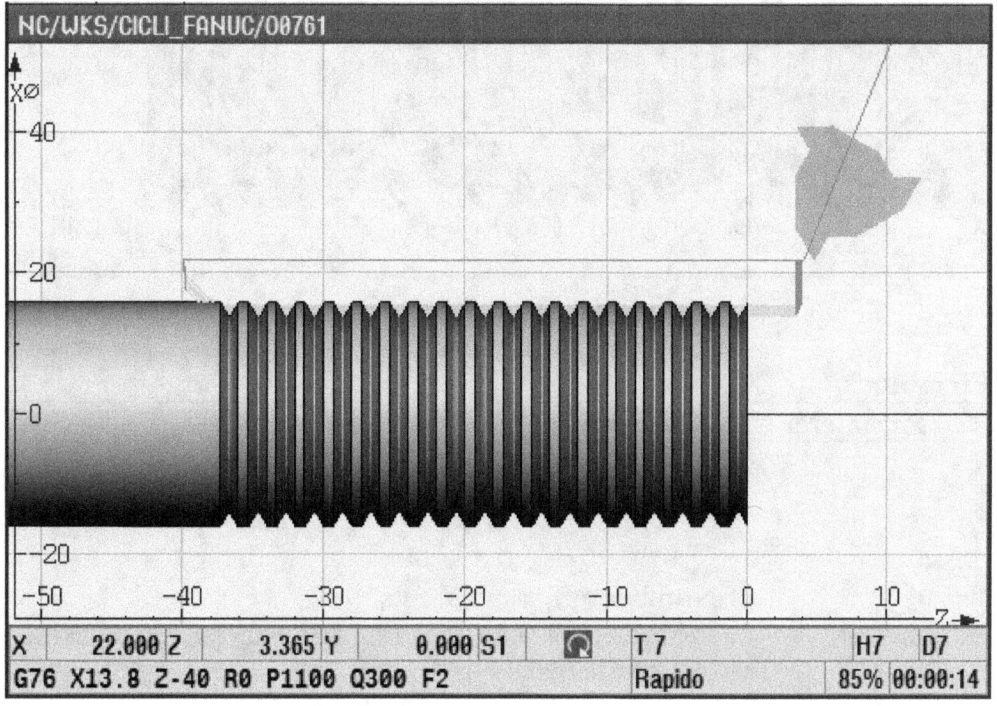

Fig. 37. G76: cycle movements

10.2 Cycle cancelation functions

This cycle is self-canceling: no further functions are necessary to deactivate it.

10.3 Cycle parameters

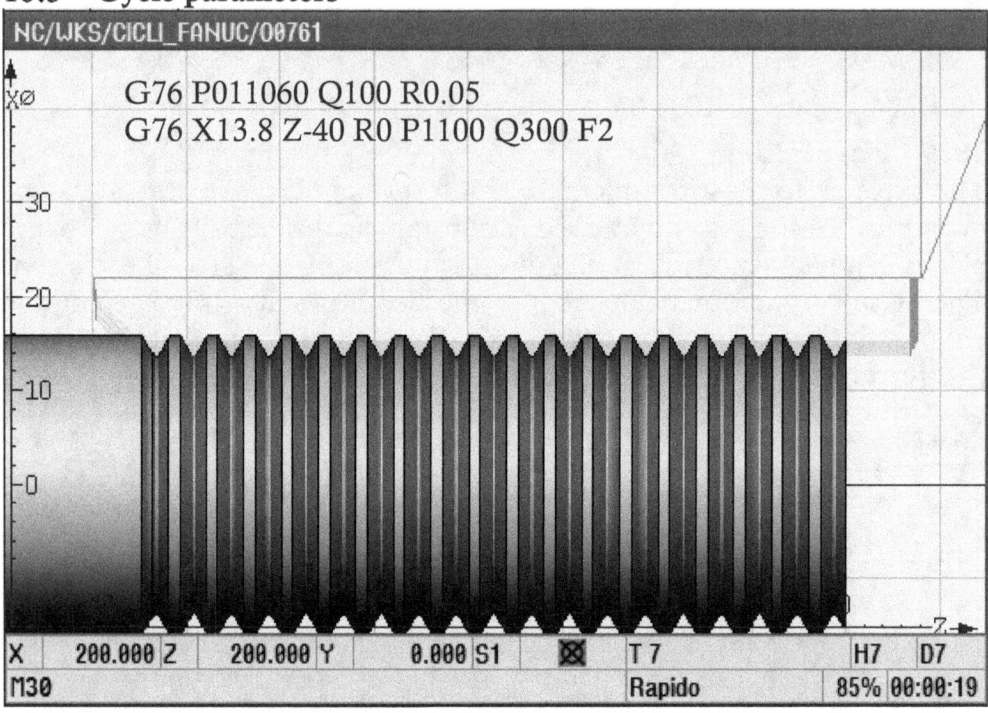

Fig. 38. G76: cycle parameters

First block

Parameter	Description
P	Six-digit data entry in three pairs.
(00) First pair	Number of finishing cuts. From 1 to 99. The finishing allowance is set by using one of the cycle parameters.

(00) Second pair	Thread exit direction. When the value is 0, the tool reaches the Z-coordinate programmed in the cycle and exits at a 90° angle. 00 = 0 * F 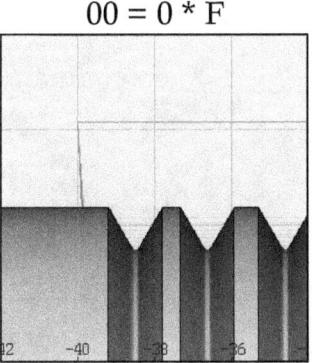 When the value is not zero (from 01 to 99), the cycle exits at a 45° angle with a quantity proportional to the lead. 10 = 1 * F 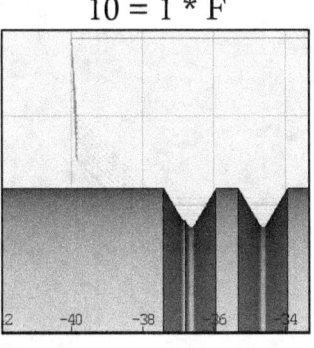 20 = 2 * F

(00) Third pair	Cutting direction. It is equal to the thread angle when the tool cuts along the side. Only 0, 29, 30, 55, 60 and 80 degrees are allowed. 00 to enter along X only, it can be used for square threads. 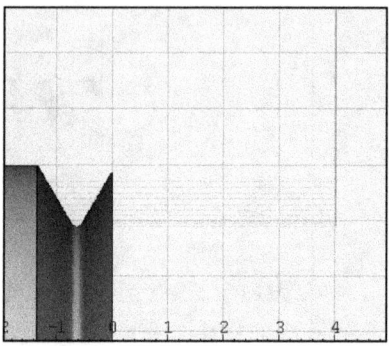 60 for metric threads 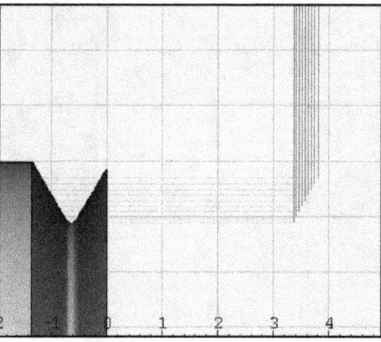 55 for Withworth threads 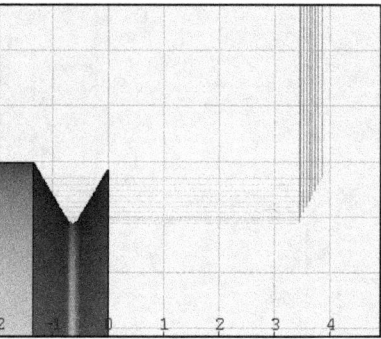

Q	Minimum depth of the cut. When this value is reached, the depth of the cut is no longer decreased. It is expressed as radial value in microns. Write 100 to set 0.1 mm. This parameter only accepts integers. If expressed in inches 1= 1/10,000 of an inch.
R	Radial stock allowance expressed in millimeters to be removed by the finishing passes. The allowance is removed in as many cuts as specified in parameter P of the first block.

Second block

Parameter	Description
X	Thread arrival diameter.
Z	Absolute coordinate of the arrival point along Z referred to the workpiece zero point.

To be continued

R	Taper value. It is equal to the starting diameter of the cut, minus the arrival diameter, divided by two. The starting and arrival diameters might not correspond to the taper diameters, but to the diameters of the cut to be performed, calculated on the basis of the tool positions in Z at the beginning and at the end of the cycle. 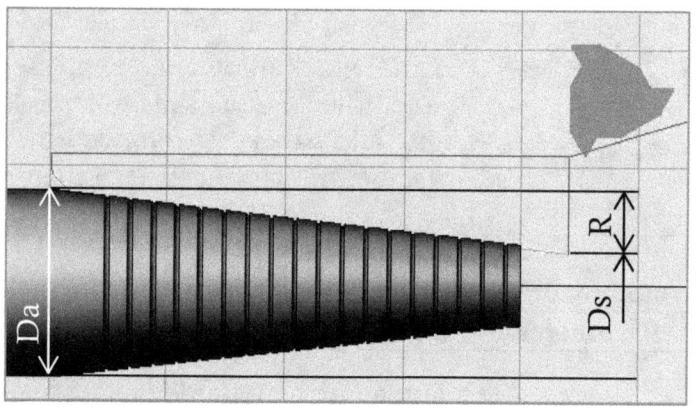 When the starting diameter is smaller than the arrival diameter, the R value is negative. When the value is equal to zero or when the parameter is not programmed, the lathe turns parallel to the Z-axis.
P	Height of the thread. This parameter is used by the cycle to determine the stock amount to be removed. Radial value expressed in microns (it only accepts integers). P is always positive. If expressed in inches 1= 1/10,000 of an inch.

Q	Depth of the first cut. Radial value expressed in microns (it only accepts integers). Q is always positive. If expressed in inches 1= 1/10,000 of an inch.
F	Thread lead. F is always positive.

10.4 Programming examples

10.4.1 Cylindrical thread

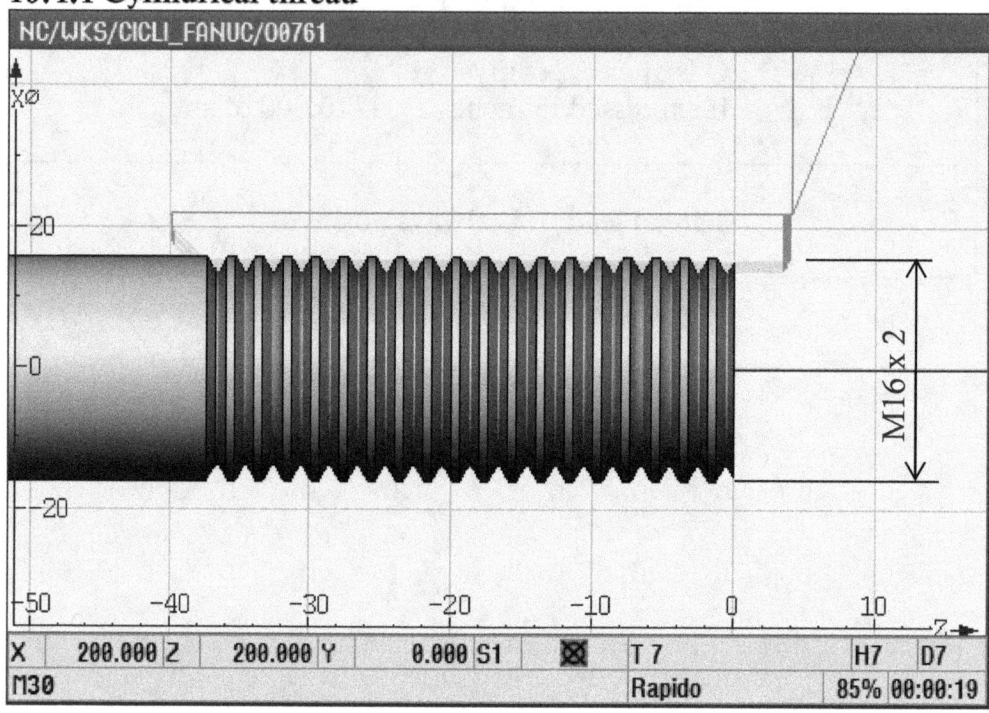

Fig. 39. G76: programming example

```
%
O0761
(G76 COMPLETE CYLINDER THREADING CUTS)
G290 ;SIEMENS LANGUAGE ACTIVATION
WORKPIECE(,,,"CYLINDER",192,0,-80,-70,16)
G291 ;FANUC LANGUAGE ACTIVATION
G18 (X-Z PLANE)
G90 (ABSOLUTE PROGRAMMING)
T0707 (EXTERNAL THREADING TOOL)
G97 S1000 M3 (SPINDLE CONSTANT REVOLUTION)
G95 (WORKING FEED RATE IN MM/REV)

G0 X22 Z4 (CYCLE START POINT)
G76 P011060 Q100 R0.05
G76 X13.8 Z-40 R0 P1100 Q300 F2
G0 X200
G0 Z200
M5
M30
%
```

10.4.2 Tapered thread

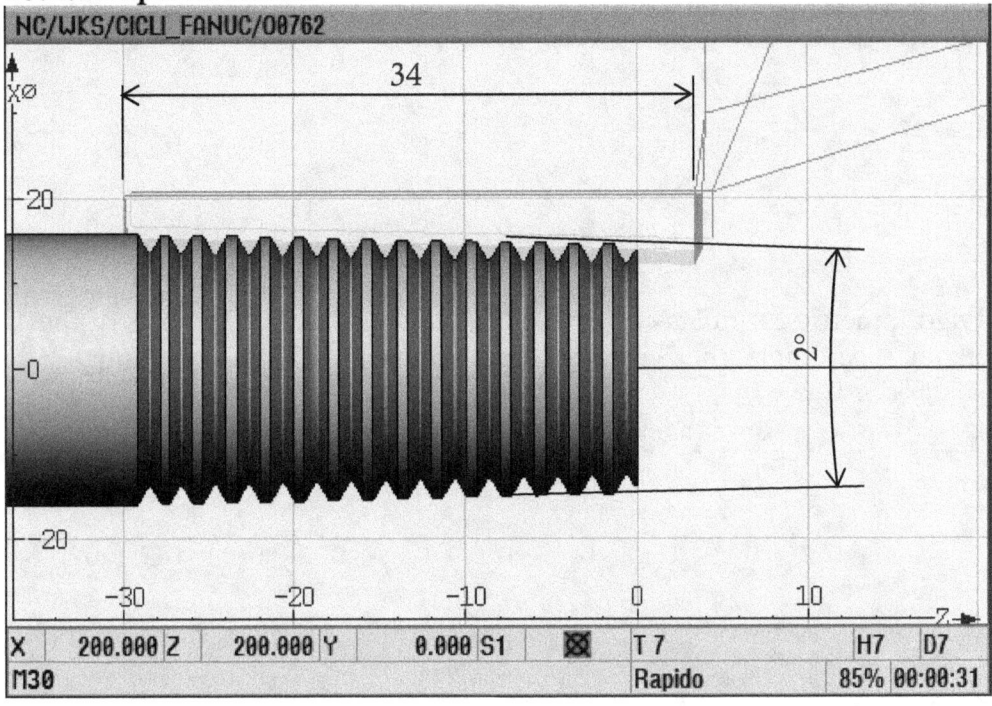

Fig. 40. G76: programming example

```
%
O0762
(G76 COMPLETE TAPER THREADING CUTS)
G290  ;SIEMENS LANGUAGE ACTIVATION
WORKPIECE(,,,"CYLINDER",192,0,-80,-70,16)
G291  ;FANUC LANGUAGE ACTIVATION
G18 (X-Z PLANE)
G90 (ABSOLUTE PROGRAMMING)

T0101 (EXTERNAL TURNING TOOL)
G97 S1000 M4 (SPINDLE CONSTANT REVOLUTION)
G0 X20 Z4 (CYCLE START POINT)
G95 (WORKING FEED RATE IN MM/REV)
```

G77 X16 Z-30 R-0.593 F0.15

```
G0 X80 Z100 (CYCLE CANCELATION)
M5 (STOP SPINDLE ROTATION)

T0707 (EXTERNAL THREADING TOOL)
G97 S1000 M3 (SPINDLE CONSTANT REVOLUTION)
G95 (WORKING FEED RATE IN MM/REV)
```

```
G0 X30 Z4 (CYCLE START POINT)
G76 P010060 Q100 R0.05
G76 X13.8 Z-30 R-0.593 P1100 Q300 F2

G0 X200
G0 Z200

M5
M30
%
```

Trigonometric formulas can be used to calculate R on the basis of the 1-degree taper angle, the Z4 start position and the Z-30 end position:

$$R = \tangent 1° * 34 = 0.0174 * 34 = 0.593$$

11. G74: grooving cycle along the Z-axis
(G74-A, G76-C)

11.1 Description

This cycle performs a complete axial groove, the stock material is removed by using one or more cuts parallel to the Z-axis, it is possible to set the cuts with or without chip breaking. This cycle cannot perform machining operations with chip removal.

G74 was designed as an axial grooving cycle, but it can also be used for axial drilling along the Z-axis or for internal or external turning cuts with chip breaking.

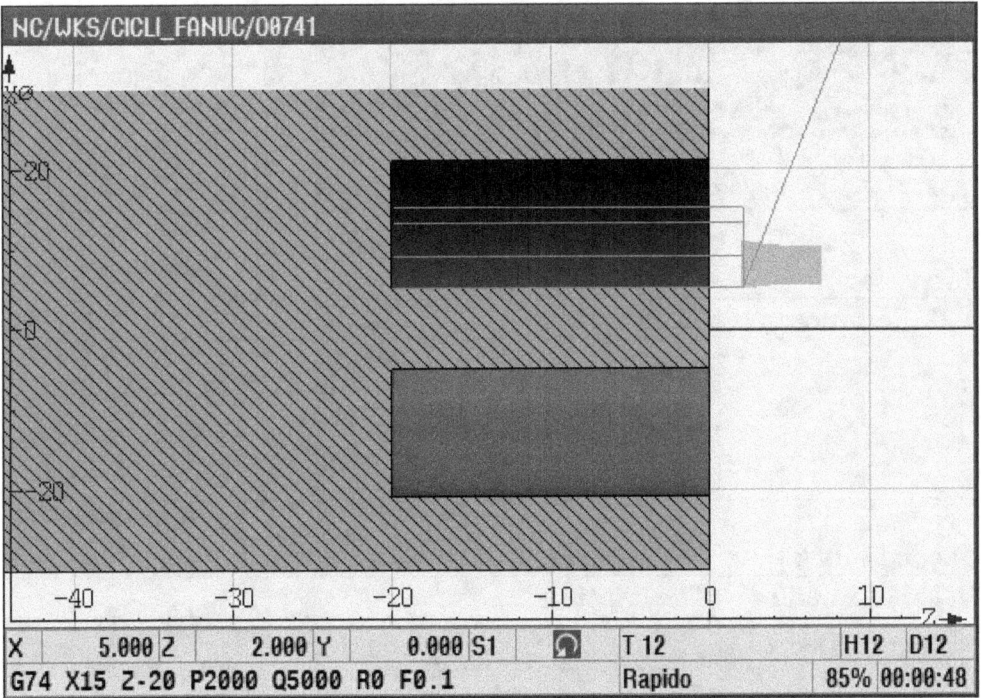

Fig. 41. G74: cycle movements

11.2 Cycle cancelation functions

This cycle is self-canceling: no further functions are necessary to deactivate it.

11.3 Cycle parameters

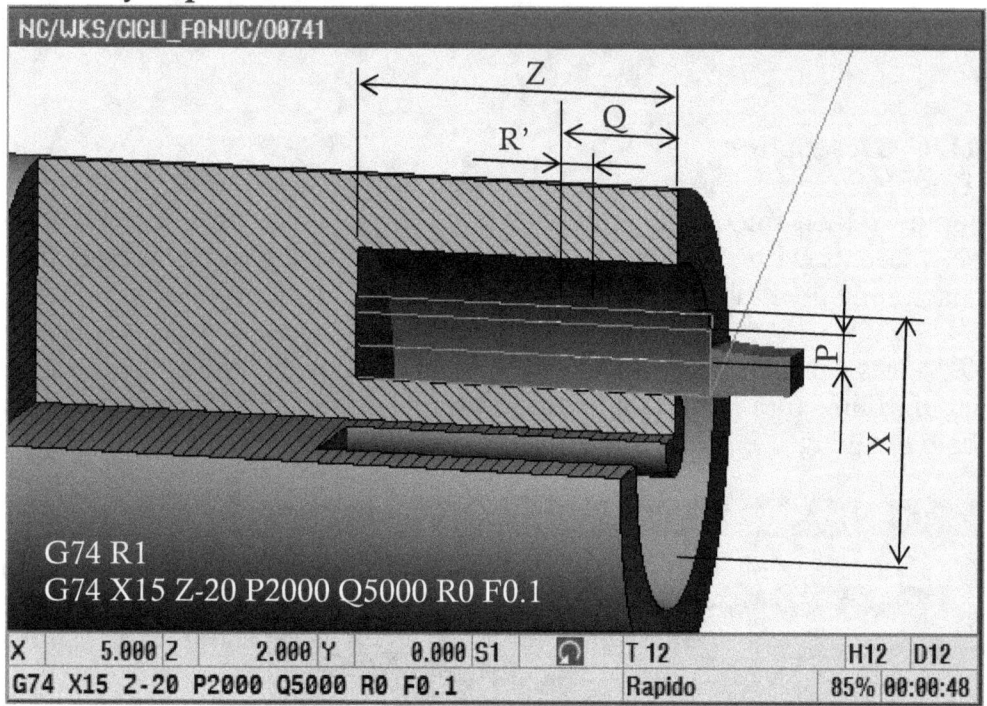

Fig. 42. G74: cycle parameters

First block

Parameter	Description
R'	Incremental return movement of the tool during chip breaking. This value is expressed in millimeters.

Second block

Parameter	Description
X	Groove arrival diameter. The starting diameter corresponds to the position programmed before the cycle.

	When it is not programmed, the grooving cycle is carried out only on the diameter programmed before the cycle (for example when it is used for drilling or turning cuts). **To set the values of the starting and arrival diameter**, it is necessary to consider the cutting edge on which the tool has been zeroed and the width of the insert.
Z	Groove (or hole, or turning cut) arrival point. The letter Z indicates the absolute coordinate (referring to the workpiece zero point) along the Z-axis of the groove end point.
P	Radial tool movement; this value determines the width of the cut. To widen grooves, 70% of the insert width is recommended. This value is expressed in thousandths of a millimeter.
Q	Cut length (along the Z-axis) before chip breaking. This value is expressed in thousandths of a millimeter.
R	Radial retraction at the end of the cut, before the tool's rapid return to the Z-point set before the cycle. **Reminder**: set this parameter to "0" when you drill axial holes. Set this parameter to "0" also when you groove a workpiece blank, otherwise the tool would interfere with the material during the return of the first cut.
F	Working feed rate.

11.4 Programming examples

11.4.1 Front groove

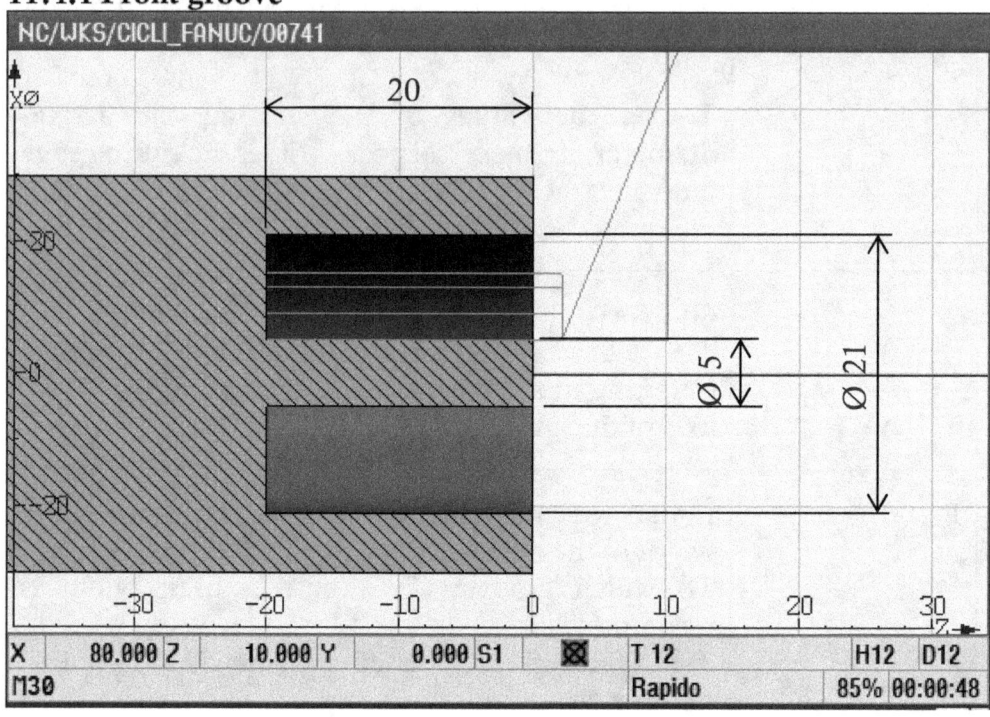

Fig. 43. G74: programming example

```
%
O0741(G74 GROOVING CYCLE ALONG Z-AXIS)
G290 ;SIEMENS LANGUAGE ACTIVATION
WORKPIECE(,,,"CYLINDER",192,0,-90,-80,30)
G291 ;FANUC LANGUAGE ACTIVATION
G18 (X-Z PLANE)
G90 (ABSOLUTE PROGRAMMING)

T1212 (FRONTAL GROOVING TOOL, INSERT WIDTH 3MM)
G97 S1400 M4 (SPINDLE CONSTANT REVOLUTION)
G95 (WORKING FEED RATE IN MM/REV)

G0 X5 Z2 (CYCLE START POINT)
G74 R1
G74 X15 Z-20 P2000 Q5000 R0 F0.1
G0 Z10
G0 X80
M5
M30
%
```

11.4.2 External turning with chip breaking

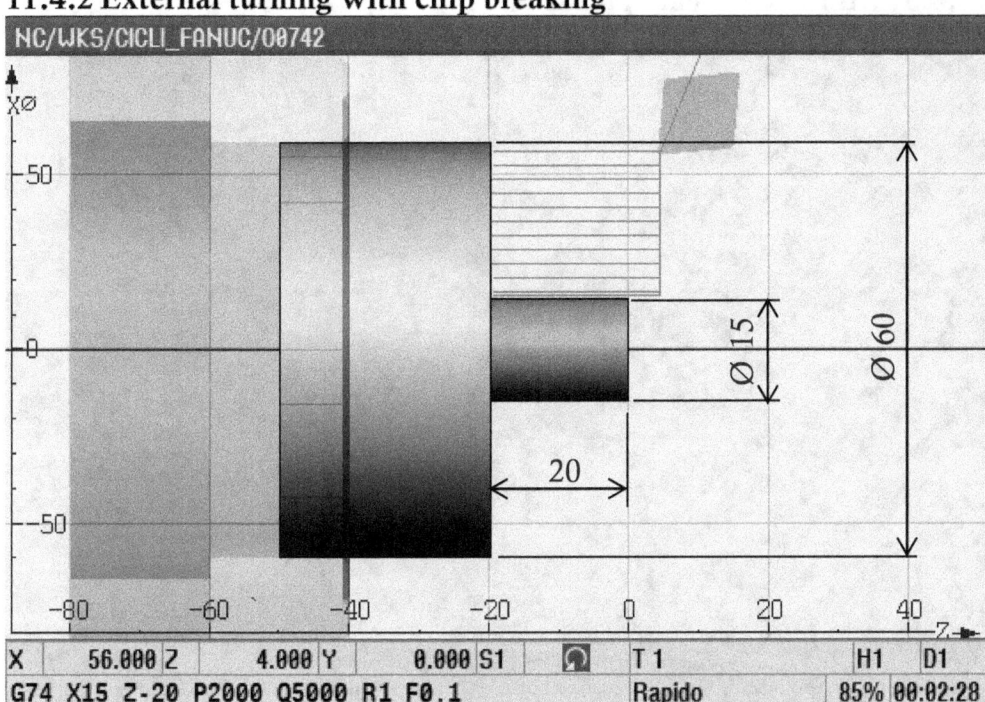

Fig. 44. G74: programming example

```
%
O0742
(G74 TURNING WITH CHIP BREAKING ALONG Z-AXIS)
G290 ;SIEMENS LANGUAGE ACTIVATION
WORKPIECE(,,,"CYLINDER",192,0,-50,-40,60)
G291 ;FANUC LANGUAGE ACTIVATION
G18 (X-Z PLANE)
G90 (ABSOLUTE PROGRAMMING)

T0101 (EXTERNAL TURNING TOOL)
G92 S3000 (SPINDLE RPM MAX LIMIT)
G96 S100 M4 (CONSTANT CUTTING SPEED)
G95 (WORKING FEED RATE IN MM/REV)

G0 X56 Z4 (CYCLE START POINT)
G74 R1
G74 X15 Z-20 P2000 Q5000 R1 F0.1
G0 Z10
G0 X80
M5
M30
%
```

11.4.3 Drilling with chip breaking

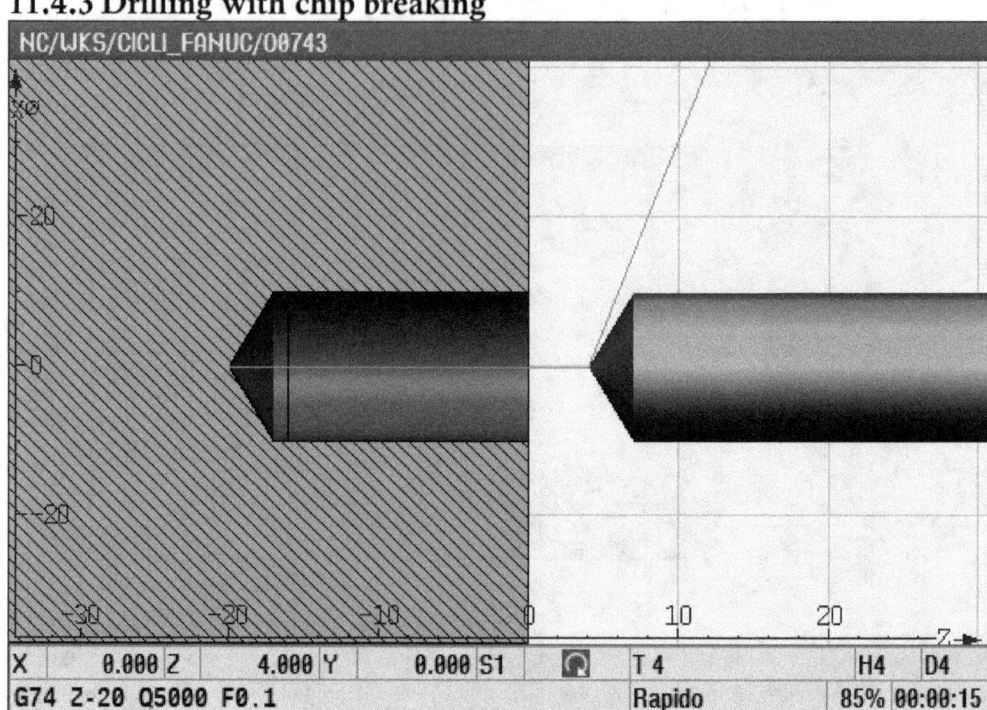

Fig. 45. G74: programming example

```
%
O0743
(G74 DRILLING WITH CHIP BREAKING ALONG Z-AXIS)
G290 ;SIEMENS LANGUAGE ACTIVATION
WORKPIECE(,,,"CYLINDER",192,0,-50,-40,60)
G291 ;FANUC LANGUAGE ACTIVATION
G18 (X-Z PLANE)
G90 (ABSOLUTE PROGRAMMING)

T0404 (AXILA DRILL D.10)
G97 S1400 M3 (SPINDLE CONSTANT REVOLUTION)
G95 (WORKING FEED RATE IN MM/REV)

G0 X0 Z4 (CYCLE START POINT)
G74 R1
G74 Z-20 Q5000 F0.1

G0 Z10
G0 X80
M5
M30
%
```

11.4.4 Boring with chip breaking

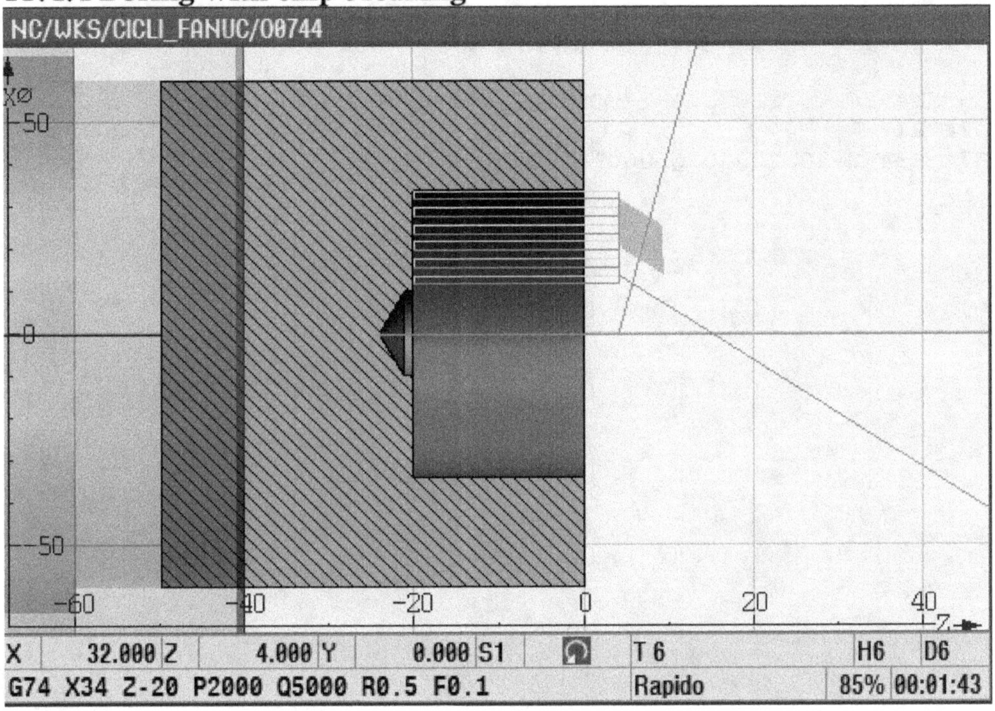

Fig. 46. G74: programming example

```
%
O0744
(G74 INTERNAL TURNING WITH CHIP BREAKING ALONG Z-AXIS)

G290 ;SIEMENS LANGUAGE ACTIVATION
WORKPIECE(,,,"CYLINDER",192,0,-50,-40,60)
G291 ;FANUC LANGUAGE ACTIVATION
G18 (X-Z PLANE)
G90 (ABSOLUTE PROGRAMMING)

T0404 (AXIAL DRILL D.10)
G97 S1400 M3 (SPINDLE CONSTANT REVOLUTION)
G95 (WORKING FEED RATE IN MM/REV)

G0 X0 Z4 (CYCLE START POINT)
G74 R1
G74 Z-24 P2000 Q5000 R0 F0.1 (X IS NOT PROGRAMMED)

G0 Z100
G0 X100
```

```
T0606 (INTERNAL TURNING TOOL)
G97 S1250 M4 (SPINDLE CONSTANT REVOLUTION)
G95 (WORKING FEED RATE IN MM/REV)

G0 X14 Z4 (CYCLE START POINT)
**G74 R1**
**G74 X34 Z-20 P2000 Q5000 R0.5 F0.1**

G0 Z200
G0 X200
M5
M30
%
```

12. G75: grooving cycle along the X-axis
(G75-A, G77-C)

12.1 Description

This cycle performs a complete axial groove, the stock material is removed by using one or more cuts parallel to the X-axis; it is possible to set the cuts with or without chip breaking. This cycle cannot perform machining operations with chip removal.

G75 was designed as a radial grooving cycle, but it can also be used for radial drilling along the X-axis or for internal or external parting-off operations with chip breaking.

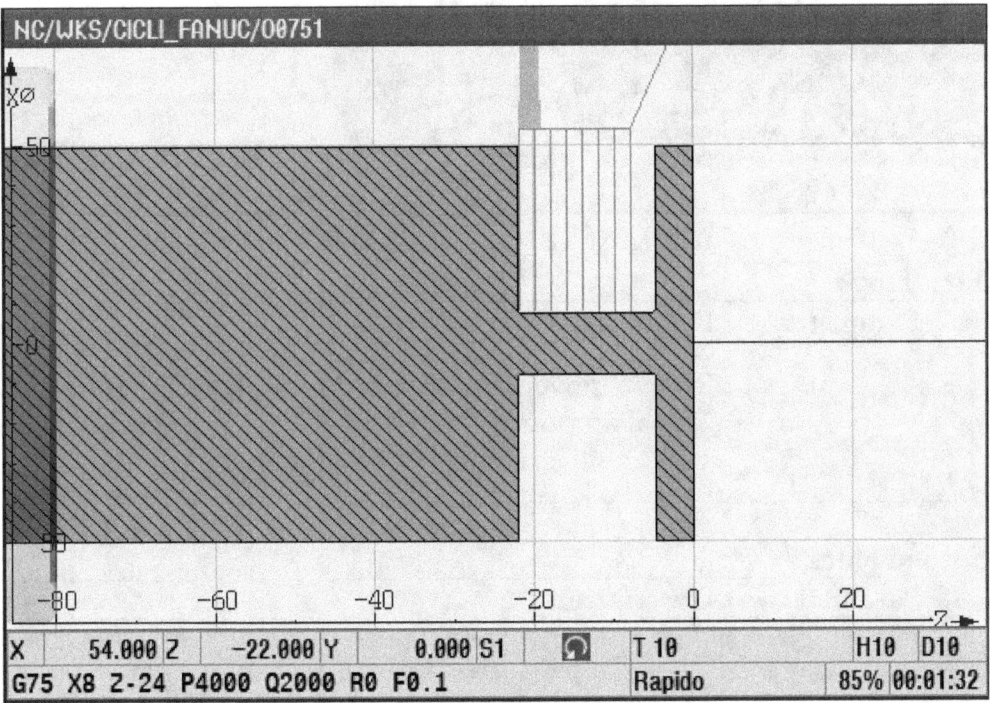

Fig. 47. G75: cycle movements

12.2 Cycle cancelation functions

This cycle is self-canceling: no further functions are necessary to deactivate it.

12.3 Cycle parameters

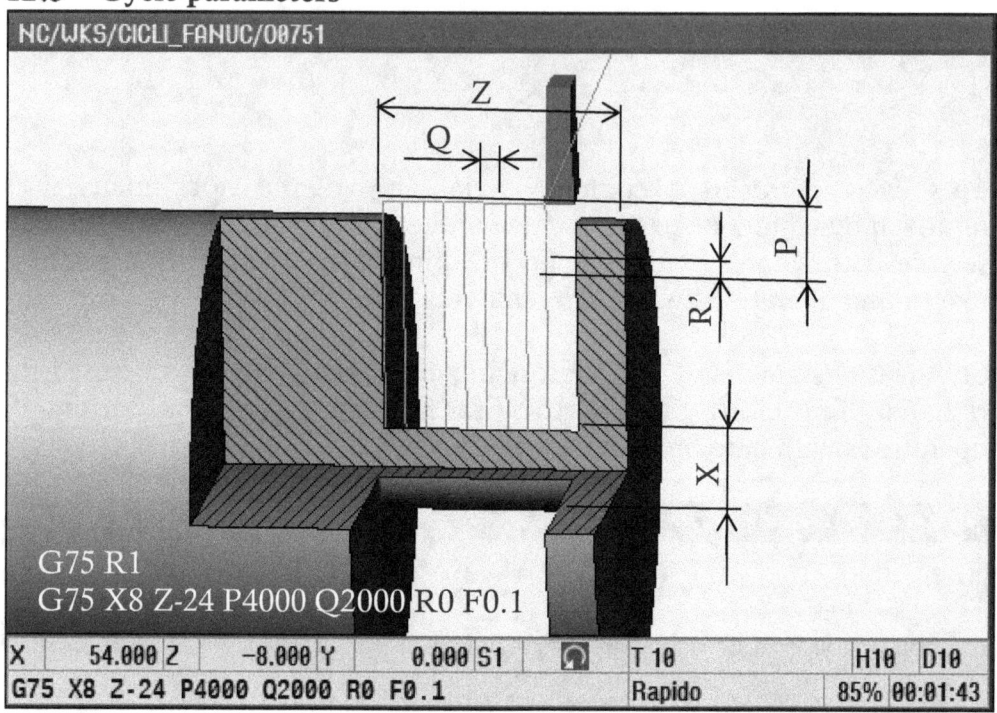

Fig. 48. G75: cycle parameters

First block

Parameter	Description
R'	Return movement of the tool during chip breaking (radial amount). This value is expressed in millimeters.

Second block

Parameter	Description
X	Groove arrival diameter. The starting diameter corresponds to the position programmed before the cycle.

Z	Groove (or hole, or parting-off) arrival point. The letter Z indicates the absolute coordinate (referring to the workpiece zero point) along the Z-axis of the groove end point. **When it is not programmed,** the grooving cycle is carried out only on the Z position programmed before the cycle (for example when it is used for drilling or parting-off). **To set values of the starting and arrival Z position**, it is necessary to consider the cutting edge on which the tool has been zeroed and the width of the insert.
P	Cut length (along the X-axis) before chip breaking (radial value). This value is expressed in thousandths of a millimeter.
Q	Longitudinal tool movement; this value determines the width of the cut. To widen grooves, 70% of the insert width is recommended. This value is expressed in thousandths of a millimeter.
R	Longitudinal retraction at the end of the cut, before the tool's rapid return movement to the X-point set before the cycle. **Reminder**: Set this parameter to "0" when you drill radial holes. Set this parameter to "0" also when you groove a workpiece blank, otherwise the tool would interfere with the material during the return of the first cut.
F	Working feed rate.

12.4 Programming example

12.4.1 Radial groove

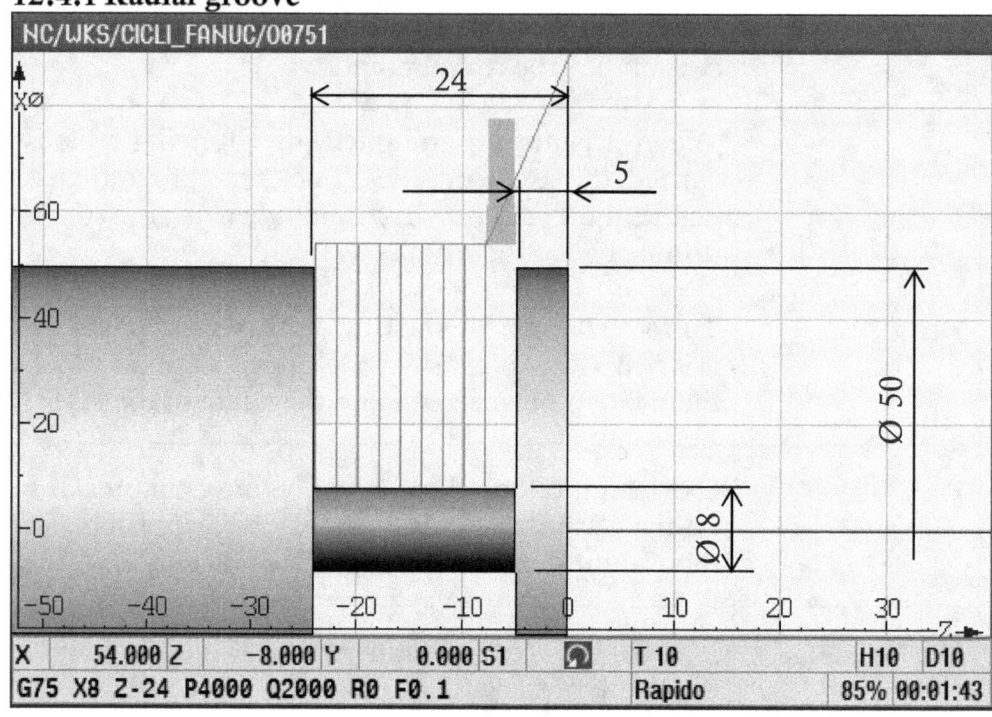

Fig. 49. G75: programming example

```
%
O0751(G75 GROOVING ALONG X-AXIS)
G290 ;SIEMENS LANGUAGE ACTIVATION
WORKPIECE(,,,"CYLINDER",192,0,-90,-80,50)
G291 ;FANUC LANGUAGE ACTIVATION
G18 G90 (X-Z PLANE, ABSOLUTE PROGRAMMING)

T1010 (RADIAL GROOVING TOOL, INSERT WIDTH 3MM)
G92 S3000 (SPINDLE RPM MAX LIMIT)
G96 S70 M4 (CONSTANT CUTTING SPEED)
G95 (WORKING FEED RATE IN MM/REV)

G0 X54 Z-8 (CYCLE START POINT)
G75 R1
G75 X8 Z-24 P4000 Q2000 R0 F0.1
G0 X200
G0 Z200
M5
M30
%
```

12.4.2 Parting-off with chip breaking

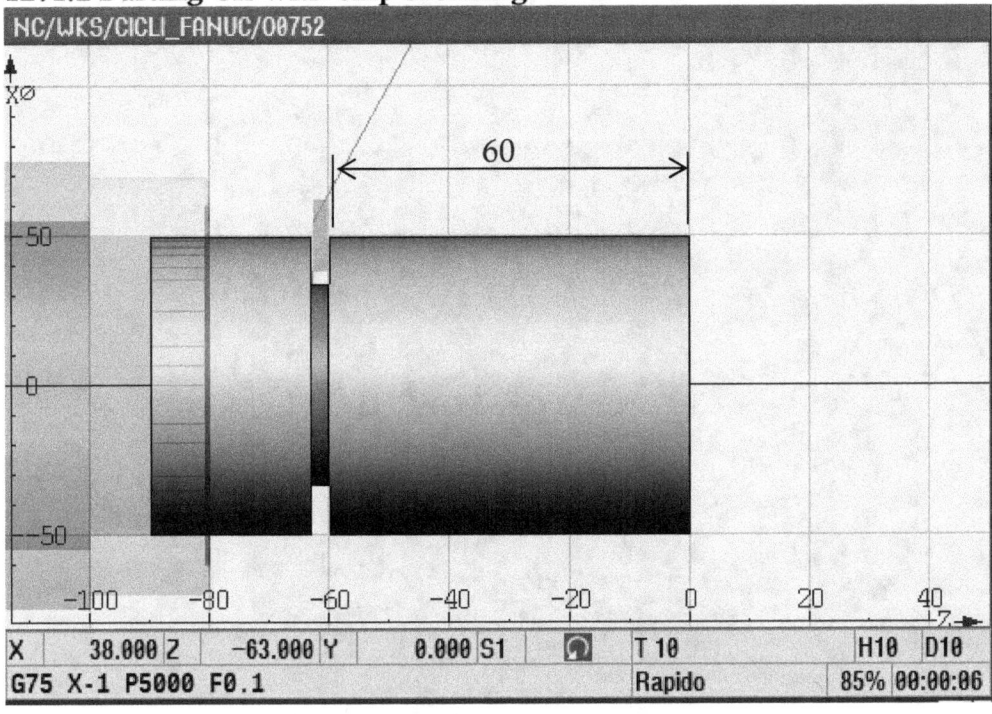

Fig. 50. G75: programming example

```
%
O0752
(G75 PARTING-OFF WITH CHIP BREAKING)
G290 ;SIEMENS LANGUAGE ACTIVATION
WORKPIECE(,,,"CYLINDER",192,0,-90,-80,50)
G291 ;FANUC LANGUAGE ACTIVATION
G18 (X-Z PLANE)
G90 (ABSOLUTE PROGRAMMING)

T1010 (RADIAL GROOVING TOOL, INSERT WIDTH 3MM)
G92 S3000 (SPINDLE RPM MAX LIMIT)
G96 S70 M4 (CONSTANT CUTTING SPEED)
G95 (WORKING FEED RATE IN MM/REV)

G0 X54 Z-63 (CYCLE START POINT)
G75 R1
G75 X-1 P5000 F0.1
G0 X200
G0 Z200
M5
M30
%
```

12.4.3 Facing with chip breaking

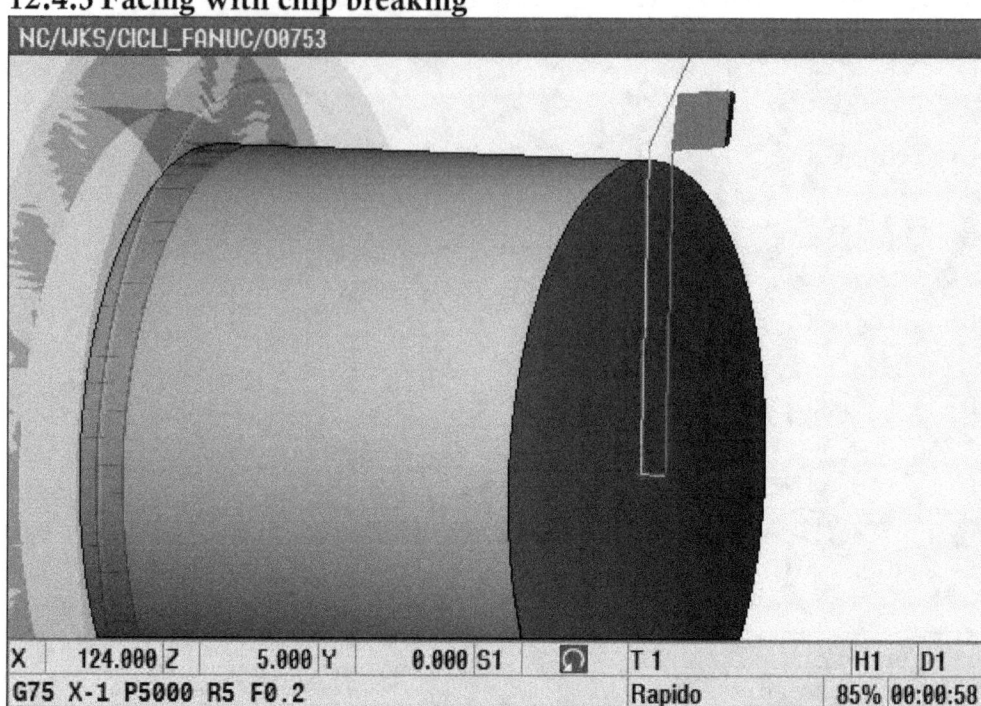

Fig. 51. G75: programming example

```
%
O0753(G75 FACING WITH CHIP BREAKING)
G290 ;SIEMENS LANGUAGE ACTIVATION
WORKPIECE(,,,"CYLINDER",192,2,-90,-80,120)
G291 ;FANUC LANGUAGE ACTIVATION
G18 (X-Z PLANE)
G90 (ABSOLUTE PROGRAMMING)

T0101 (TURNING TOOL)
G92 S3000 (SPINDLE RPM MAX LIMIT)
G96 S100 M4 (CONSTANT CUTTING SPEED)
G95 (WORKING FEED RATE IN MM/REV)

G0 X124 Z0 (CYCLE START POINT)
G75 R1
G75 X-1 P5000 R5 F0.2
G0 X200
G0 Z200
M5
M30
%
```

12.4.4 Radial drilling with chip breaking

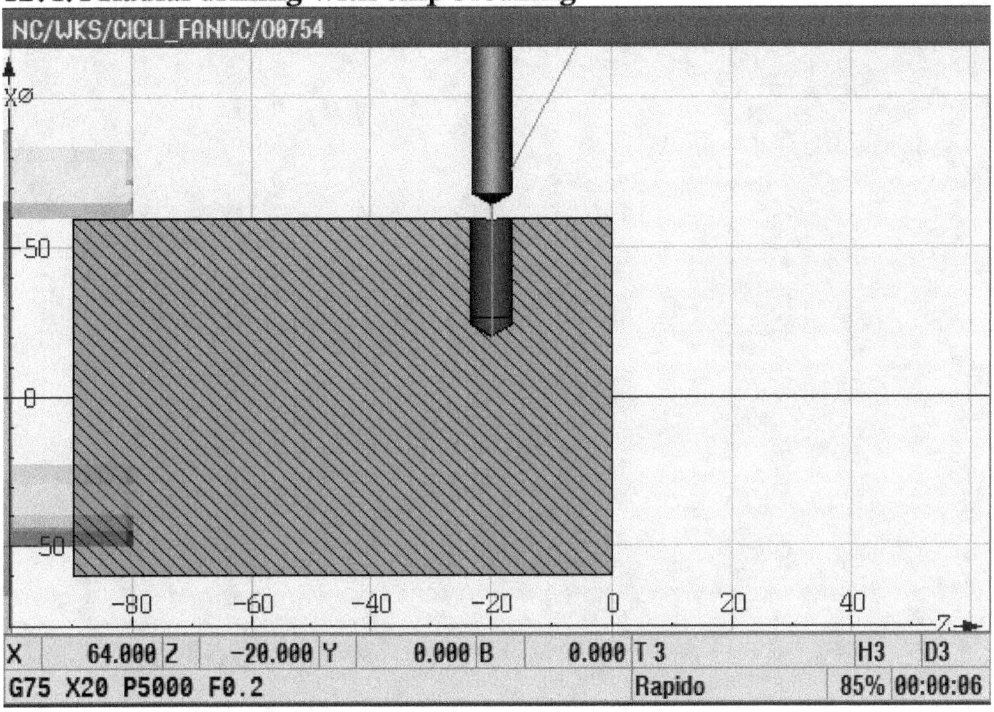

Fig. 52. G75: programming example

```
%
O0754
(G75 RADIAL DRILLING WITH CHIP BREAKING)
G290 ;SIEMENS LANGUAGE ACTIVATION
WORKPIECE(,,,"CYLINDER",192,0,-90,-80,60)
G291 ;FANUC LANGUAGE ACTIVATION
G18 (X-Z PLANE)
G90 (ABSOLUTE PROGRAMMING)

T0303 (RADIAL DRILL D.6.8)
G290 ;ENABLE SIEMENS LANGUAGE TO SELECT THE DRIVEN TOOL AS
MAIN SPINDLE
SETMS(3)
G97 S2000 M3 ;TOOL ROTATION, CONSTANT RPM
G291 ;FANUC LANGUAGE ACTIVATION

M19 B0 (MAIN SPINDLE ANGULAR ORIENTATION)

G0 X64 Z-20 (CYCLE START POINT)
G75 R0.5
G75 X20 P5000 F0.2
```

98

```
G0 X200
G0 Z200

M5
M30
%
```

13. G83: drilling cycle along the Z-axis
(G83-A, G83-C)

13.1 Description

This cycle performs drilling operations with chip breaking or chip removal; if no cut depth is specified by parameter Q, drilling is carried out with only one cut.

The cycle carries out the following movements.
1. The cycle starts from the point programmed before the cycle.

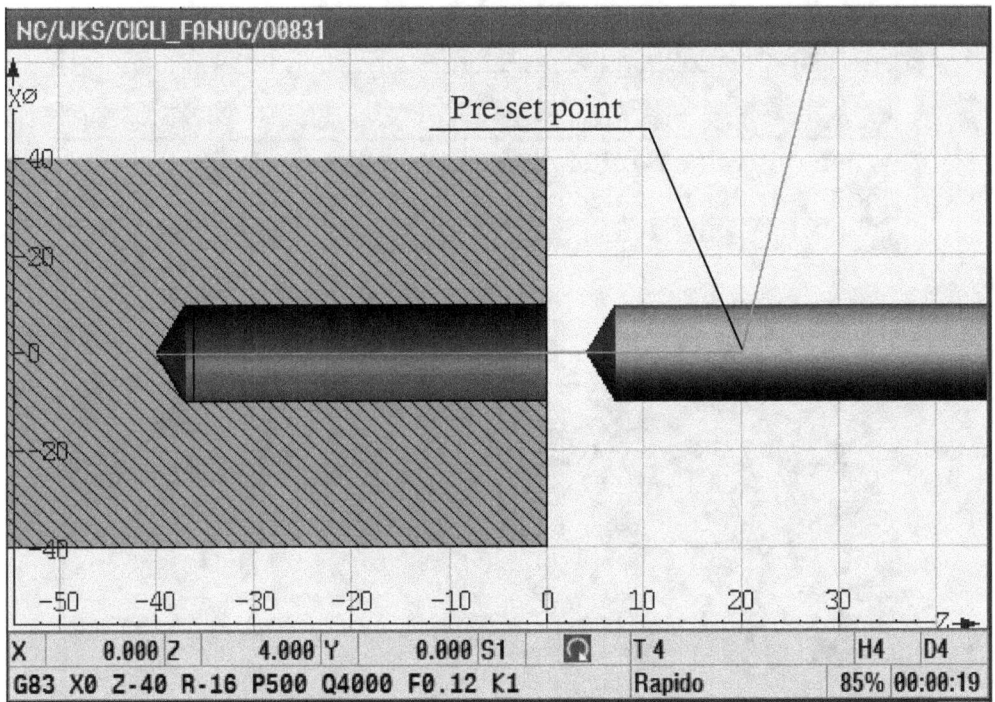

Fig. 53. G83: cycle movements

2   Rapid movement to the "X-coordinate" programmed in the cycle. This parameter defines the diametrical position of the hole. The value is "0" when the hole is on the main axis, the drill is fixed and the workpiece rotates.

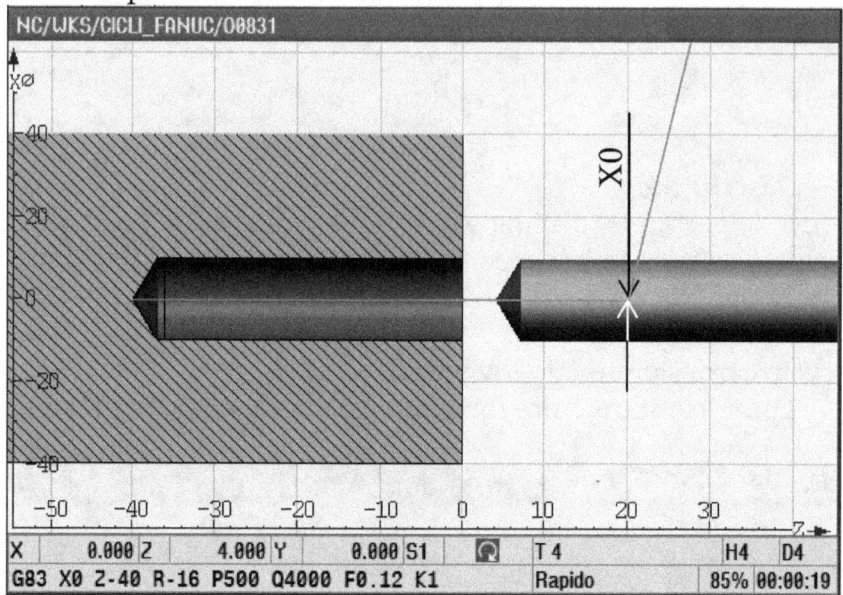

When this value is different from zero, the workpiece is stationary and the drill is mounted on a driven holder.

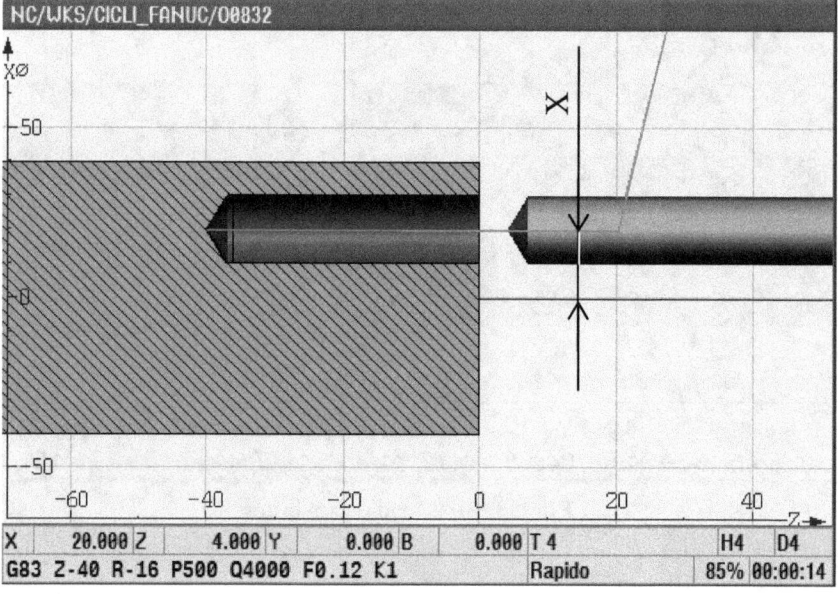

When the parameter "X" is not programmed, drilling is carried out at the point set before the cycle.

3   Rapid movement to the incremental distance defined in the parameter "R". When the parameter is not programmed, drilling is carried out at the point set before the cycle.

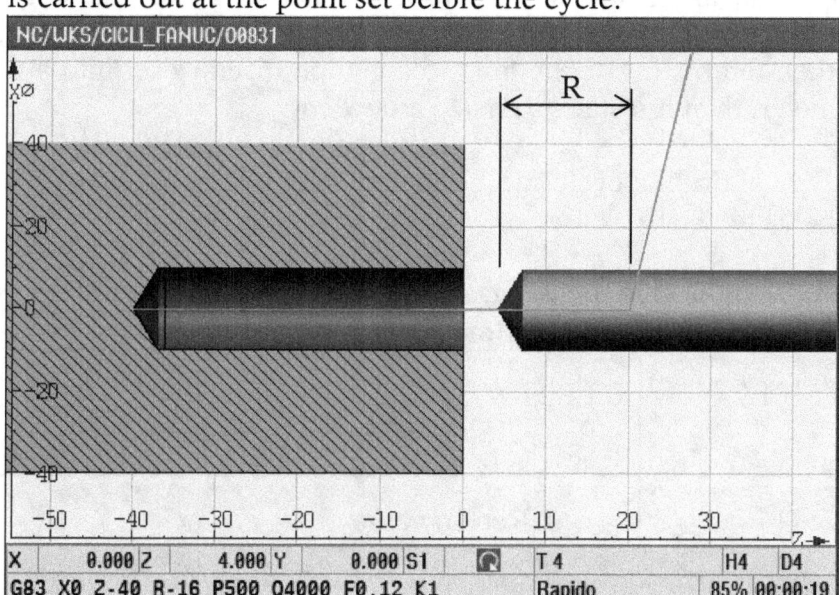

4   Drilling is performed until the cycle reaches the "Z-coordinate" programmed in the cycle. Chip breaking or chip removal is carried out on the basis of the NC parameter settings.

5   Once the drill reaches the end of the hole, the tool stops for the programmed dwell time, then it is rapidly retracted, first to the point defined in parameter "R", then to the point programmed before the cycle.

13.2 Cycle-related NC parameters

"Chip breaking" means that the tool starts cutting over the length set in the cycle, then it moves backwards of a certain distance set into a machine parameter, then it restarts to make a new cut.

"Chip removal" means that the tool starts cutting over the length set in the cycle, then it gets out of the hole till the point defined by parameter R and then returns inside the hole to a certain distance set into a machine parameter, then it restarts to make a new cut.

The type of drilling cycle can be set by using two parameters: N. 5101.2 and N. 5114.

To program the chip breaking, set the parameters as follows:

Parameter	Description
N. 5101.2	Bit RTR = 0 **Chip breaking** is performed
N. 5114	= ... Return value

To program the chip removal, set the parameters as follows:

Parameter	Description
N. 5101.2	Bit RTR = 1 **Chip removal** is performed
N. 5114	= ... Clearance value

13.3 Cycle cancelation functions

The function is modal and repeats the drilling operation for each programmed position after the cycle activation, its deactivation is done with G80.

13.4 Cycle parameters

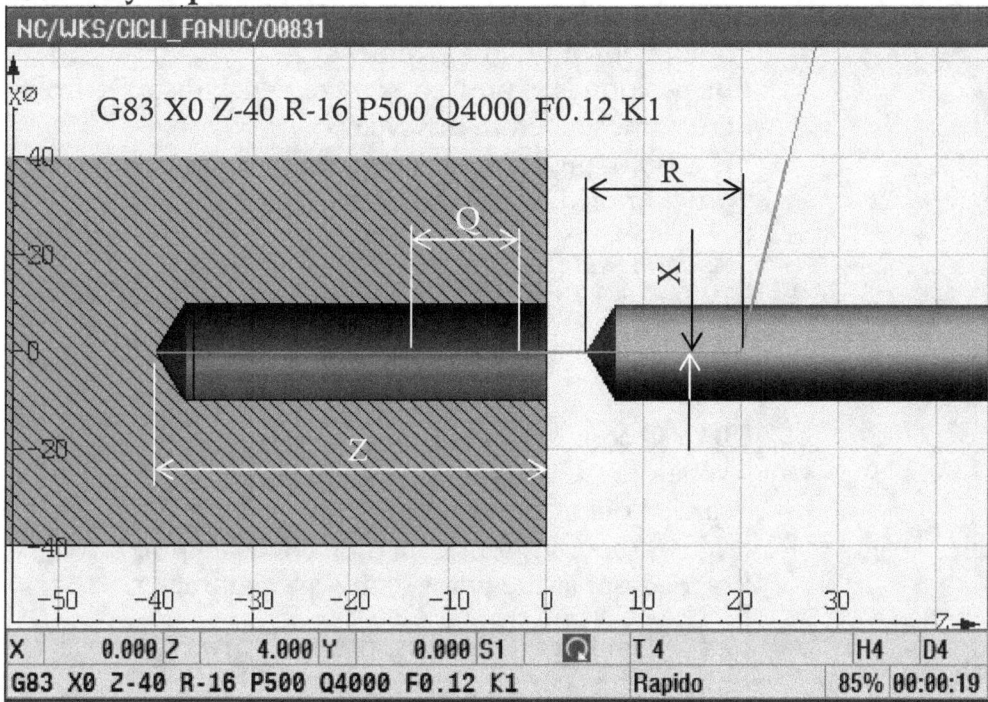

Fig. 54. G83: cycle parameters

Parameter	Description
X	X-coordinate of the cycle start point. If it is not programmed, the position in X remains the position of the point set before the cycle.
C	Potential angular position of the workpiece. When it is programmed in the cycle, the C-axis should be enabled before recalling the cycle. Otherwise, the workpiece should be oriented angularly before the cycle.
Z	Absolute coordinate at the bottom of the hole along the Z-axis.

R	Drilling start point along Z. The cycle reaches this point with a rapid movement. Its position is expressed as incremental distance from the point set before the cycle. If it is not programmed, drilling starts from the point programmed before the cycle. **A-code system** When chip removal is performed, the tool always returns to the point set before the cycle, regardless of this point. **B or C code system** When the G94 function is enabled, chip removal is carried out at the point set before the cycle. When the G95 function is enabled, chip removal is carried out at the point set by this parameter. At the end of the process, the cycle always returns to this point before rapidly returning to the point programmed before the cycle.
Q	Length of cut (expressed in micron).
P	Dwell time at the bottom of the hole (expressed in milliseconds).
F	Working feed rate.

K	Number of cycle repetitions. This parameter is used to drill multiple front holes. 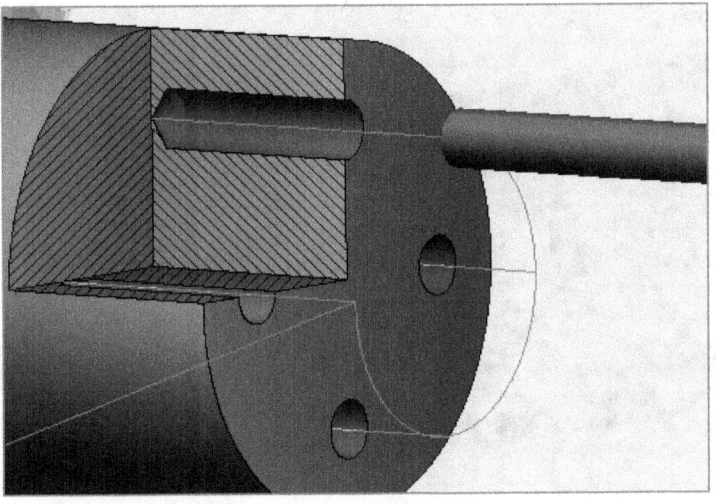 Program the functions for the activation of the "C-axis" before the cycle. This parameter needs to be programmed together with the incremental distance between the holes. For example, program "H90" and "K4" to cut four holes at a 90° angle on the "C-axis". If K0 is selected, no hole is made. After the angular orientation, some machines require brake activation to lock the spindle. Hence, program the M function according to the manufacturer's instructions (e.g.: H90 K4 M31).

13.5 Programming example

13.5.1 Axial drilling

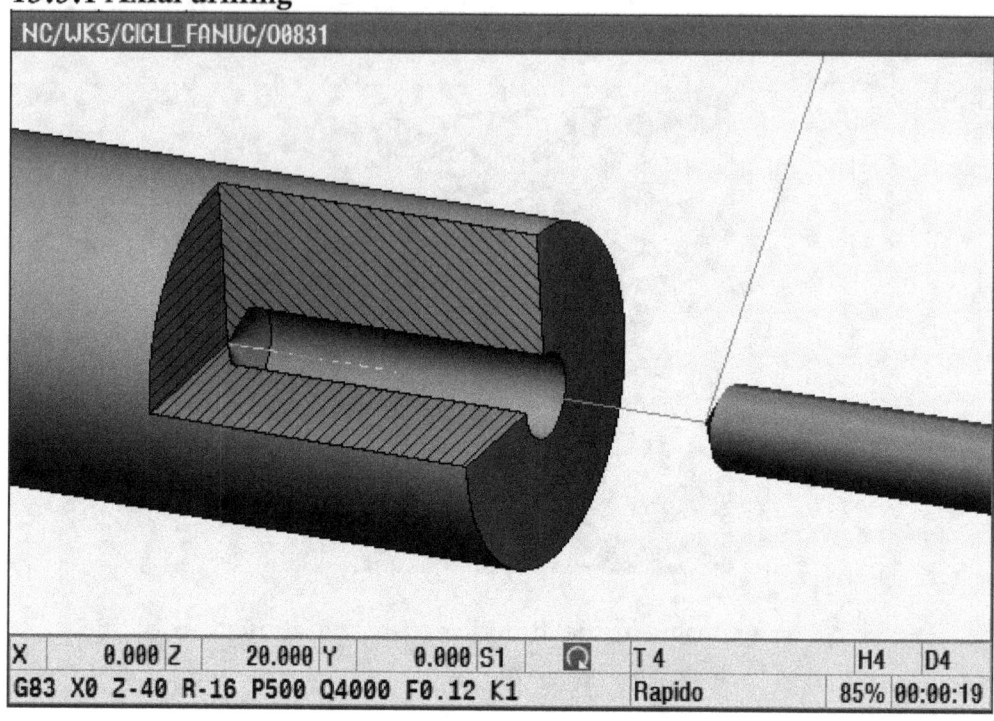

Fig. 55. G83: programming example

```
%
O0831
(G83 AXIAL DRILLING WITH CHIP REMOVAL)
G290 ;SIEMENS LANGUAGE ACTIVATION
WORKPIECE(,,,"CYLINDER",192,0,-90,-80,40)
G291 ;FANUC LANGUAGE ACTIVATION
G18 (X-Z PLANE)
G90 (ABSOLUTE PROGRAMMING)

T0404 (AXIAL DRILL D.10)
G97 S1400 M3 (SPINDLE CONSTANT REVOLUTION)
G95 (WORKING FEED RATE IN MM/REV)

G0 X0 Z20 (CYCLE START POINT)
G83 X0 Z-40 R-16 P500 Q4000 F0.12 K1
G80 (CYCLE CANCELATION)
G0 X200 Z200
M5
M30
%
```

14. G87: drilling cycle along the X-axis
(G87-A, G87-C)

14.1 Description

This cycle performs drilling operations with chip breaking or chip removal; if no depth of cut is specified by parameter Q, drilling is carried out with only one cut.

The cycle carries out the following movements.
1   The cycle starts from the point programmed before the cycle.

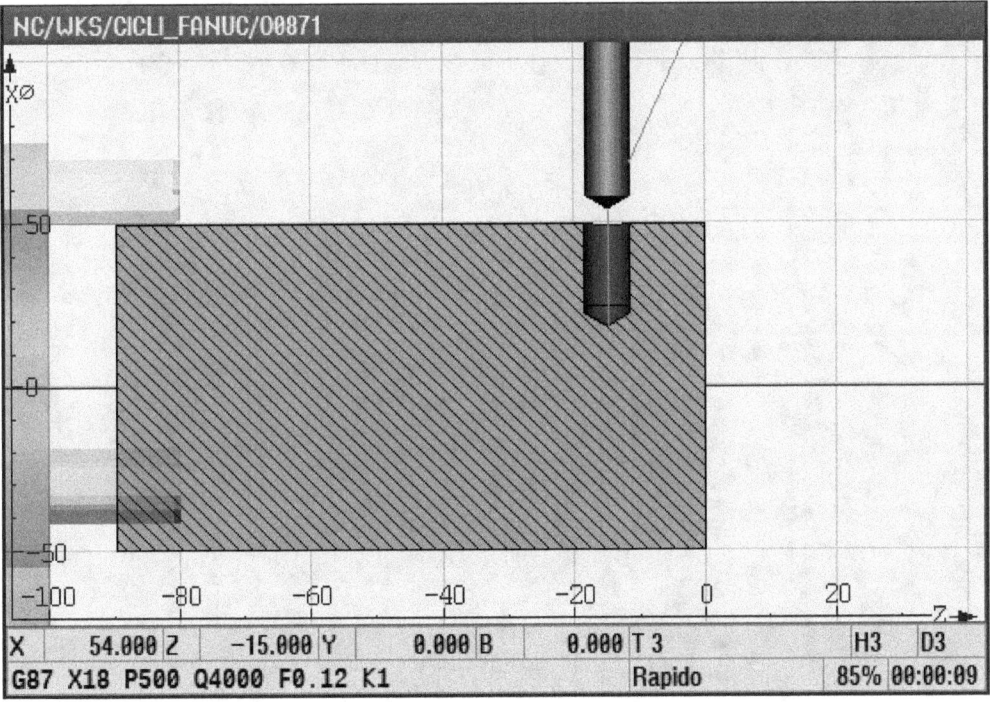

Fig. 56. G87: cycle movements

2   Rapid movement to the "Z-coordinate" programmed in the cycle. This parameter defines the Z drilling position. When the parameter "Z" is not programmed, drilling is performed at the point set before the cycle.

3   Rapid movement to the incremental distance defined in the parameter "R". When the parameter is not programmed, drilling is performed from the point set before the cycle.

4   Drilling is performed until the cycle reaches the "X-coordinate" programmed in the cycle. Chip breaking or chip removal is carried out on the basis of the NC parameter settings.

5   Once the drill reaches the end of the hole, the tool stops for the programmed dwell time, then it is rapidly retracted, first to the point defined in parameter "R", then to the point programmed before the cycle.

14.2 Cycle-related NC parameters

"Chip breaking" means that the tool starts cutting over the length set in the cycle, then it moves backwards of a certain distance set into a machine parameter, then it restarts to make a new cut.

"Chip removal" means that the tool starts cutting over the length set in the cycle, then it gets out of the hole till the point defined by parameter R and then returns inside the hole to a certain distance set into a machine parameter, then it restarts to make a new cut.

The type of drilling cycle can be set by using two parameters: N. 5101.2 and N. 5114.

To program the chip breaking, set the parameters as follows:

Parameter	Description
N. 5101.2	Bit RTR = 0 **Chip breaking** is performed
N. 5114	= ... Return value

To program the chip removal, set the parameters as follows:

Parameter	Description
N. 5101.2	Bit RTR = 1 **Chip removal** is performed
N. 5114	= ... Clearance value

14.3 Cycle cancelation functions
Use function G80 to cancel this cycle.

14.4 Cycle parameters

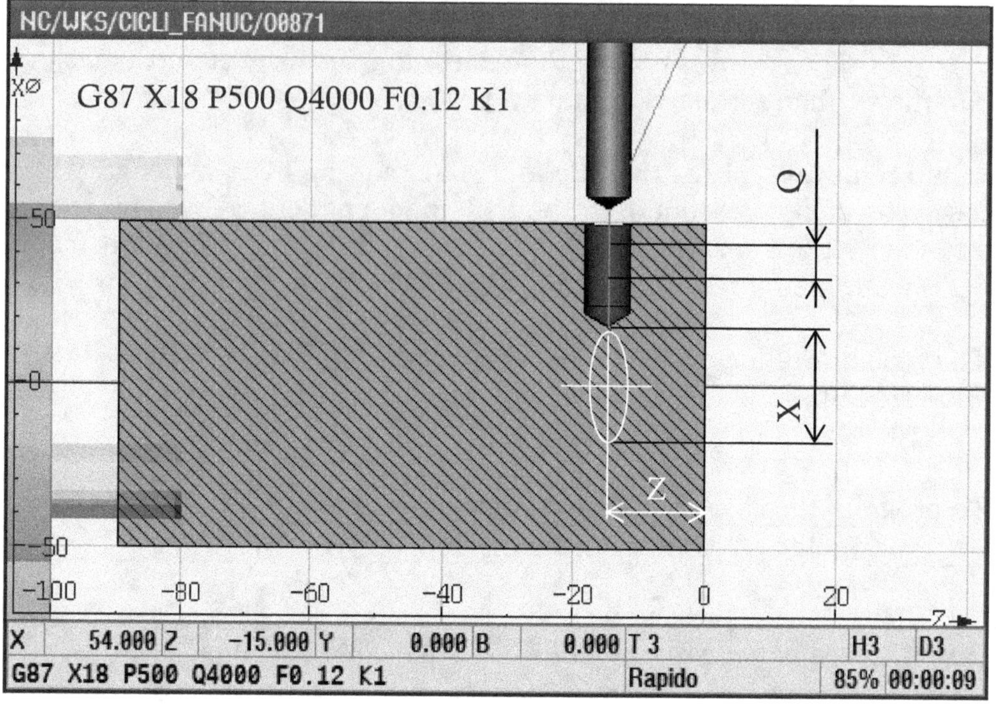

Fig. 57. G87: cycle parameters

Parameter	Description
Z	Z-coordinate of the cycle start point. If it is not programmed, the position in Z remains the position of the point set before the cycle.
C	Potential angular position of the workpiece. When it is programmed in the cycle, the C-axis should be enabled before recalling the cycle. Otherwise, the workpiece should be oriented angularly before the cycle
X	Absolute coordinate at the bottom of the hole along the X-axis.

| R | Drilling start point. The cycle performs a rapid movement to R.
Its position is expressed as incremental distance from the point set before the cycle (radial value).
If it is not programmed, drilling starts from the point programmed before the cycle.

A-code system
When chip removal is performed, the tool always returns to the point set before the cycle, regardless of this point.

B or C code system
When the G94 function is enabled, chip removal is carried out at the point set before the cycle.
When the G95 function is enabled, chip removal is carried out at the point set by this parameter.

At the end of the process, the cycle always returns to this point before rapidly returning to the point programmed before the cycle. |
|---|---|
| Q | Length of cut (radial value expressed in micron). |
| P | Dwell time at the bottom of the hole (expressed in milliseconds). |
| F | Working feed rate. |

K	Number of cycle repetitions. This parameter needs to be programmed together with the incremental distance between the holes. Example 1: program "W-10" and "K2" to cut two holes at a 10 mm distance along the Z-axis. Example 2: program "H90" and "K4" to cut four holes at a 90° angle along the "C-axis". Program the functions for the activation of the "C-axis" before the cycle. If K0 is selected, no hole is made. After the angular orientation, some machines require brake activation to lock the spindle. Hence, program the M function according to the manufacturer's instructions (e.g.: H90 K4 M31).

14.5 Programming examples

14.5.1 Radial drilling

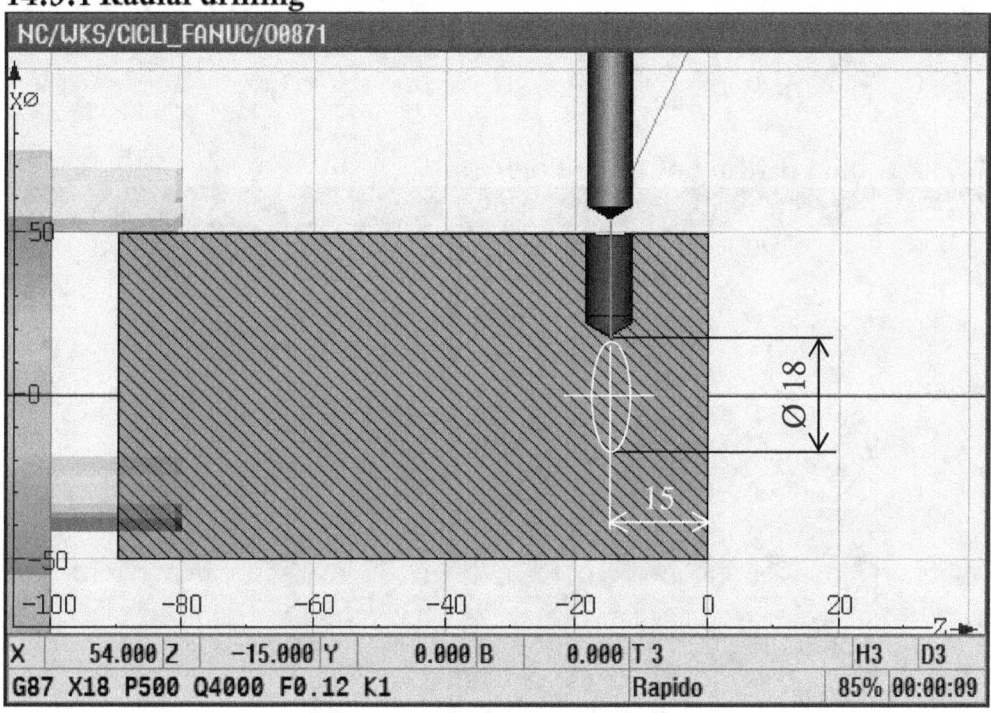

Fig. 58. G87: programming example

```
%
O0871
(G87 RADIAL DRILLING WITH CHIP REMOVAL)
G290  ;SIEMENS LANGUAGE ACTIVATION
WORKPIECE(,,,"CYLINDER",192,0,-90,-80,50)
G291  ;FANUC LANGUAGE ACTIVATION
G18 (X-Z PLANE)
G90 (ABSOLUTE PROGRAMMING)


T0303 (RADIAL DRILL D.6.8)
G290  ;ENABLE SIEMENS LANGUAGE TO SELECT THE DRIVEN TOOL AS
MAIN SPINDLE
SETMS(3)
G291  ;FANUC LANGUAGE ACTIVATION
G97 S1400 M3 (TOOL ROTATION, CONSTANT RPM)
G95 (WORKING FEED RATE IN MM/REV)

G0 X54 Z-15 (CYCLE START POINT)

M19 B0 (MAIN SPINDLE ANGULAR ORIENTATION)
```

```
G87 X18 P500 Q4000 F0.12 K1
G80 (CYCLE CANCELATION)

G0 X200 Z200
M5
M30
%
```

14.5.2 Radial drilling of three holes

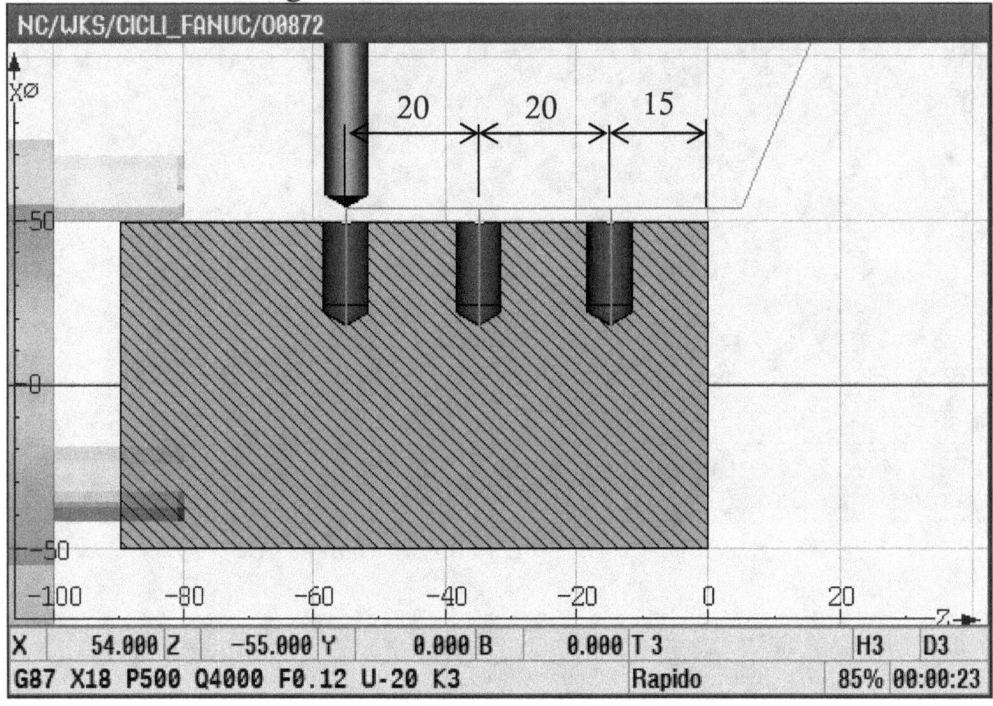

Fig. 59. G87: programming example

```
%
O0872
(G87 RADIAL DRILLING OF THREE HOLES WITH CHIP REMOVAL)
G290 ;SIEMENS LANGUAGE ACTIVATION
WORKPIECE(,,,"CYLINDER",192,0,-90,-80,50)
G291 ;FANUC LANGUAGE ACTIVATION
G18 (X-Z PLANE)
G90 (ABSOLUTE PROGRAMMING)

T0303 (DIAMETER 6.8 FOR DRILLING)
G290 ;ENABLE SIEMENS LANGUAGE TO SELECT THE DRIVEN TOOL AS
MAIN SPINDLE
SETMS(3)
G291 ;FANUC LANGUAGE ACTIVATION
```

```
G97 S1400 M3 (DRIVEN TOOL ROTATION)
G95 (WORKING FEED RATE IN MM/REV)

G0 X54 Z5 (CYCLE START POINT)

M19 B0 (MAIN SPINDLE ANGULAR ORIENTATION)

G87 X18 P500 Q4000 F0.12 W-20 K3
G80 (CYCLE CANCELATION)

G0 X200 Z200
M5
M30
%
```

14.5.3 Parting-off

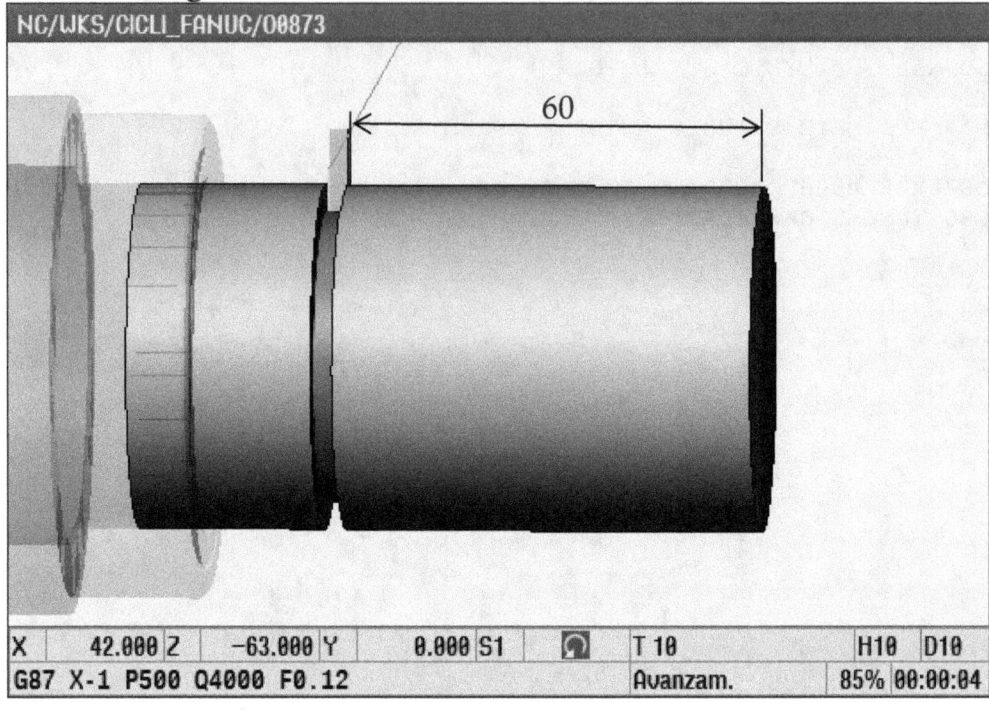

Fig. 60. G87: programming example

```
%
O0873
(G87 PARTING-OFF WITH CHIP REMOVAL)
G290 ;SIEMENS LANGUAGE ACTIVATION
WORKPIECE(,,,"CYLINDER",192,0,-90,-80,50)
G291 ;FANUC LANGUAGE ACTIVATION
G18 (X-Z PLANE)
G90 (ABSOLUTE PROGRAMMING)

T1010 (PARTING-OFF TOOL, INSTERT WIDTH 3MM)
G92 S3000 (SPINDLE RPM MAX LIMIT)
G96 S70 M4 (CONSTANT CUTTING SPEED)
G95 (WORKING FEED RATE IN MM/REV)

G0 X54 Z-63 (CYCLE START POINT)
G87 X-1 P500 Q4000 F0.12
G80 (CYCLE CANCELATION)

G0 X200 Z200
M5
M30
%
```

14.5.4 Three radial grooves

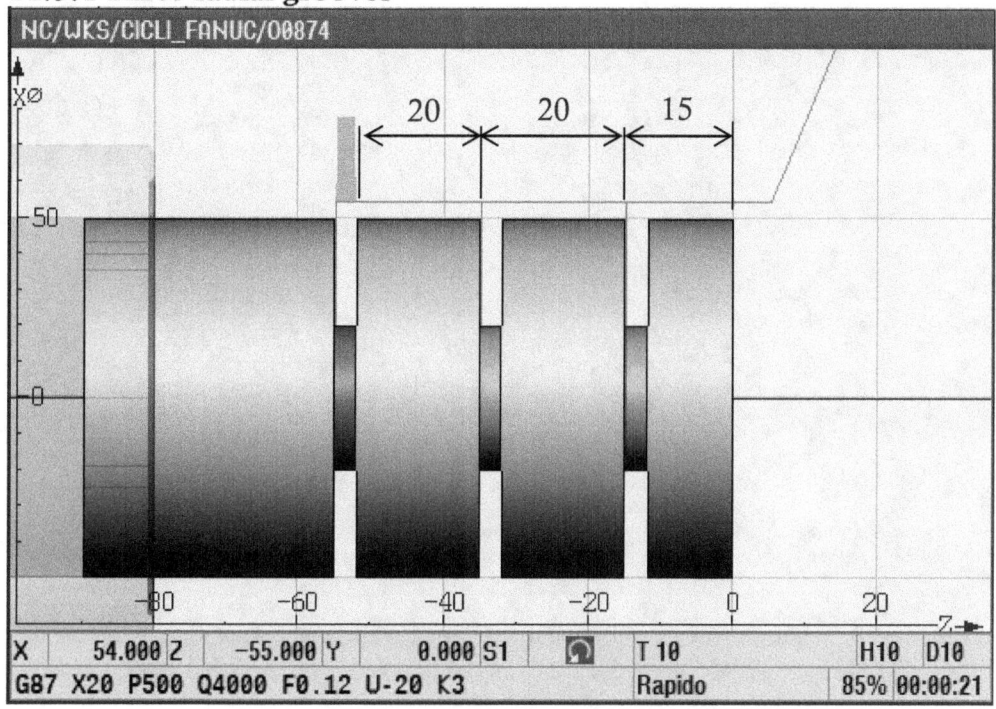

Fig. 61. G87: programming example

```
%
O0874
(G87 MULTI-GROOVE CUTTING WITH CHIP REMOVAL)
G290 ;SIEMENS LANGUAGE ACTIVATION
WORKPIECE(,,,"CYLINDER",192,0,-90,-80,50)
G291 ;FANUC LANGUAGE ACTIVATION
G18 (X-Z PLANE)
G90 (ABSOLUTE PROGRAMMING)

T1010 (PARTING-OFF TOOL, INSERT WIDTH 3MM)
G97 S1400 M4 (SPINDLE CONSTANT REVOLUTION)
G95 (WORKING FEED RATE IN MM/REV)

G0 X54 Z5 (CYCLE START POINT)
G87 X20 P500 Q4000 F0.12 W-20 K3
G80 (CYCLE CANCELATION)

G0 X200 Z200
M5
M30
%
```

15. G84: tapping cycle along the Z-axis
(G84-A, G84-C)

15.1 Description

This cycle performs a complete axial tapping by working up to the programmed coordinate, stops at the bottom of the hole for the programmed dwell time and then reverses the spindle rotation to return on the cycle start point.

The cycle can be used both for standard tapping with axial spring holder and for rigid tapping.

To perform a rigid tapping, it is necessary to program before G84 the function "M29" followed by "S" and the number of revolutions at which the machining has to be performed.

The cycle carries out the following movements.
1   The cycle starts from the point programmed before the cycle.

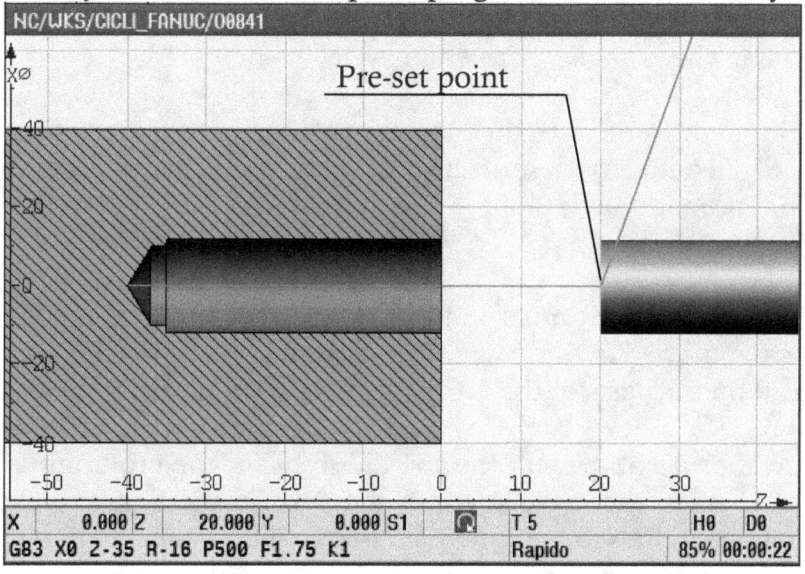

Fig. 62. G84: cycle movements

2 Rapid movement to the "X-coordinate" programmed in the cycle. This parameter defines the diametrical position of the hole. The value is "0" when the hole is on the main axis, the tapping tool is fixed and the workpiece rotates.

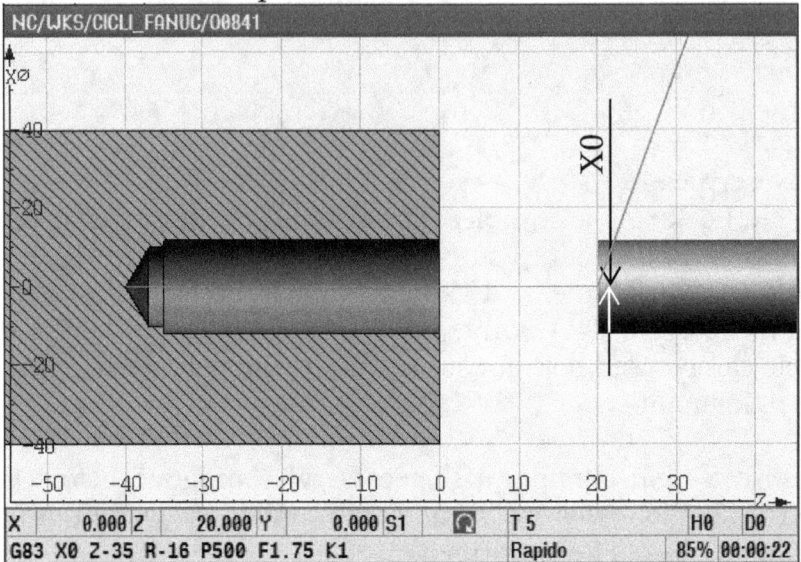

When this value is different from zero, the workpiece is stationary and the tapping tool is mounted on a driven holder.
When the parameter "X" is not programmed, tapping is carried out at the point set before the cycle.

3 Rapid movement to the incremental distance defined in parameter "R" (as already described in cycle G83). When the parameter is not programmed, tapping is carried out at the point set before the cycle.

4 Tapping is performed until the cycle reaches the "Z-coordinate" programmed in the cycle. Chip breaking or chip removal is carried out on the basis of the NC parameter settings.

5 At the end of the cut, the spindle stops for the programmed dwell time, then the cycle reverses the spindle rotation and the tap is retracted, first to point defined in parameter "R", then to the point programmed before the cycle.

15.2 Cycle-related NC parameters

To perform right- or left-hand tapping, it is necessary to reverse the spindle rotation. In many cases, the forward rotation of the tap depends on the last programmed rotation direction. Otherwise, the values of the following parameters have to be changed manually.

Parameter 5112 sets the forward rotation of the spindle.
Parameter 5113 sets the reverse rotation of the spindle.

Parameter	Description
N. 5112	= 3 clockwise spindle forward rotation = 4 counterclockwise spindle forward rotation
N. 5113	= 3 clockwise spindle reverse rotation = 4 counterclockwise spindle reverse rotation

15.3 Axial rigid peck tapping

To enable rigid peck tapping, set the following parameter:
5200 bit 5 =1. For multichannel machines, edit this parameter in the channel where the tapping is performed.

Parameter	Description
N. 5200 bit 5 (PCP)	= 1 After setting each peck by using parameter "Q" (microns), the tap fully retracts to remove the chip. = 0 After setting each peck by using parameter "Q" (microns), the tap does not fully retract; it stops and then performs the second tapping pass, continuing in this way until it reaches the arrival coordinate.

Programming example:
G84 Z-30 P500 **Q5000** F1 K1

15.4 Cycle cancelation functions

The function is modal and repeats the tapping operation for each programmed position after the cycle activation, its deactivation is done with G80.

15.5 Cycle parameters

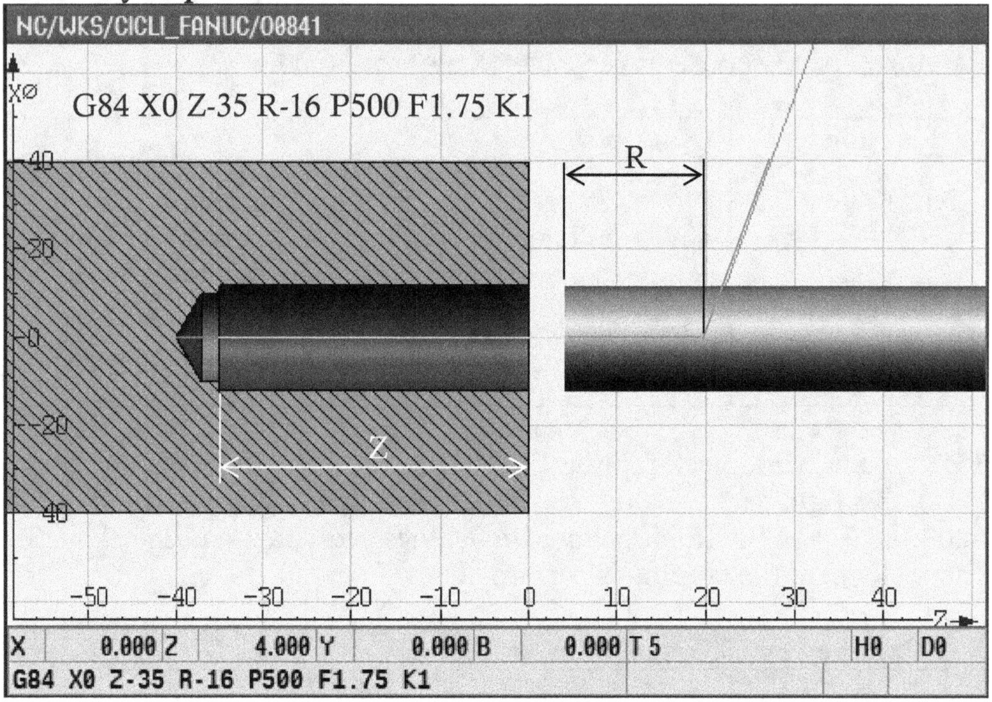

Fig. 63. G84: cycle parameters

Parameter	Description
X	X-coordinate of the cycle start point. If it is not programmed, the position in X remains the position of the cycle start point.
C	Potential angular position of the workpiece. When it is programmed in the cycle, the C-axis should be enabled before recalling the cycle. Otherwise, the workpiece should be oriented angularly before the cycle.

Z	Absolute Z-coordinate at the end of tapping.
R	Tapping start point along Z. The cycle reaches this point with a rapid movement. Its position is expressed as incremental distance from the point set before the cycle. If it is not programmed, drilling starts from the point programmed before the cycle.
P	Dwell time at the bottom of the hole expressed in milliseconds.
F	Lead of the tap
K	Number of cycle repetitions. This parameter is used to tap multiple front holes. Program the functions for the activation of the "C-axis" before the cycle.

	This parameter needs to be programmed together with the incremental distance between the holes. For example, program "H90" and "K4" to tap four holes at a 90° angle on the "C-axis". If K0 is selected, no tapping is made. After the angular orientation, some machines require brake activation to lock the spindle. Hence, program the M function according to the manufacturer's instructions (e.g.: H90 K4 M31).

15.6 Programming example

15.6.1 Axial tapping

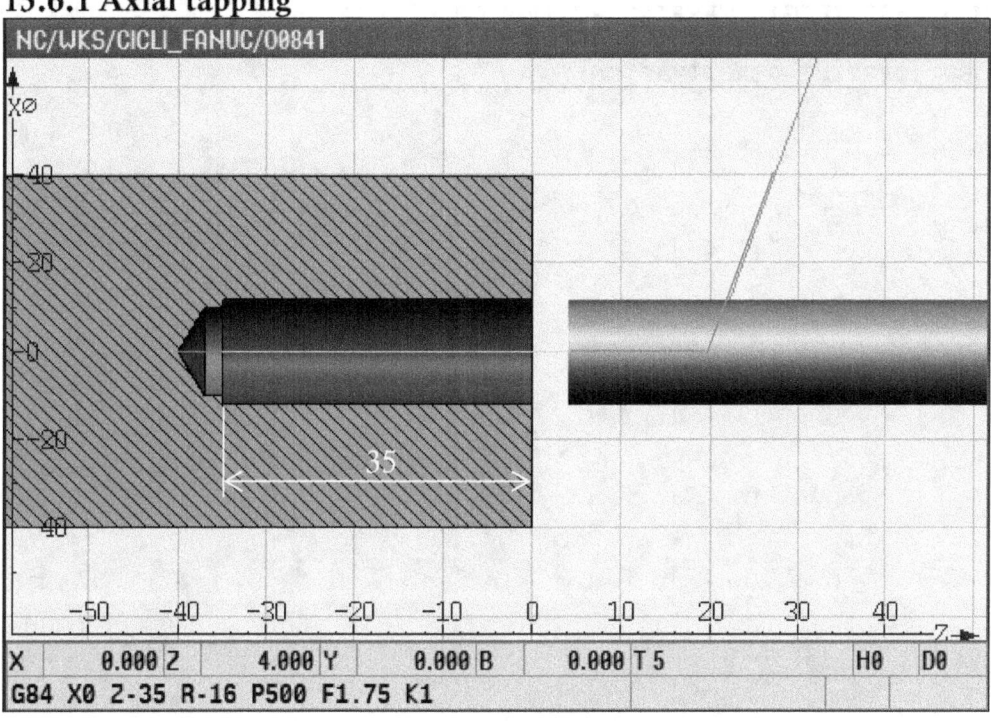

Fig. 64. G84: programming example

```
%
O0841
(G84 AXIAL RIGID TAPPING M12 F1.75)
G290 ;SIEMENS LANGUAGE ACTIVATION
WORKPIECE(,,,"CYLINDER",192,0,-90,-80,40)
G291 ;FANUC LANGUAGE ACTIVATION
G18 (X-Z PLANE)
G90 (ABSOLUTE PROGRAMMING)

T0404 (AXIAL DRILL D.10)
G97 S1400 M3 (SPINDLE CONSTANT REVOLUTION)
G95 (WORKING FEED RATE IN MM/REV)

G0 X0 Z20 (CYCLE START POINT)
G83 X0 Z-40 R-16 P500 Q4000 F0.12 K1
G80 (CYCLE CANCELATION)
G0 Z100

T0909 (AXIAL TAPPING TOOL M12 F1.75)
G97 S800 (SPINDLE CONSTANT REVOLUTION)
```

```
G95 (WORKING FEED RATE IN MM/REV)

G0 X0 Z20 (CYCLE START POINT)
M29 S800 (RIGID TAPPING ACTIVATION)
G84 X0 Z-35 R-16 P500 F1.75 K1
G80 (TAPPING DEACTIVATION)

G0 Z200 X200

M5
M30
%
```

16. G88: tapping cycle along the X-axis
(G88-A, G88-C)

16.1 Description

This cycle performs a complete radial tapping by working up to the programmed coordinate, stops at the bottom of the hole for the programmed dwell time and then reverses the spindle rotation to return on the cycle start point.
The cycle can be used both for standard tapping with axial spring holder and for rigid tapping.

To perform a rigid tapping, it is necessary to program before G88 the function "M29" followed by "S" and the number of revolutions at which the machining has to be performed.

The cycle carries out the following movements.
   1   The cycle starts from the point programmed before the cycle.

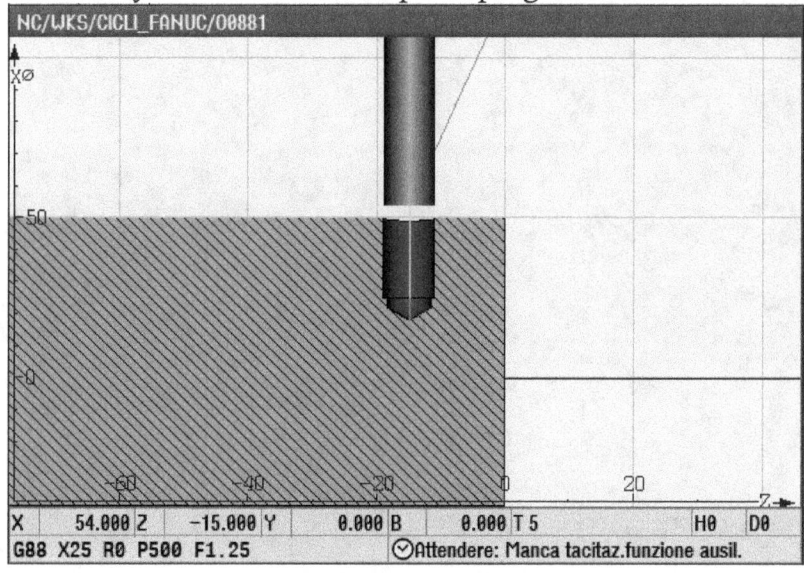

Fig. 65. G88: cycle movements

CNC – Fanuc turning cycles

2   Rapid movement to the "Z-coordinate" programmed in the cycle. This parameter defines the Z tapping position. When the parameter is not programmed, tapping is performed at the point set before the cycle.

3   Rapid movement to the point defined in parameter R. When the parameter is not programmed, tapping is performed at the point set before the cycle.

4   Tapping is performed until the cycle reaches the "X-coordinate" programmed in the cycle.

5   At the end of the cut, the tap stops for the programmed dwell time, then the cycle reverses the spindle rotation and the tap is retracted, first to point defined in parameter "R", then to the point programmed before the cycle.

16.2 Cycle-related NC parameters

To perform right- or left-hand tapping, it is necessary to reverse the spindle rotation. In many cases, the forward rotation of the tap depends on the last programmed rotation direction. Otherwise, the values of the following parameters have to be changed manually.

Parameter 5112 sets the forward rotation of the spindle.
Parameter 5113 sets the reverse rotation of the spindle.

Parameter	Description
N. 5112	= 3 clockwise spindle forward rotation = 4 counterclockwise spindle forward rotation
N. 5113	= 3 clockwise spindle reverse rotation = 4 counterclockwise spindle reverse rotation

16.3 Axial rigid peck tapping

To enable the rigid peck tapping, set the following parameter:
5200 bit 5 =1.
For multichannel machines, edit this parameter in the channel where the tapping is performed.

Parameter	Description
N. 5200 bit 5 (PCP)	= 1 After setting each peck by using parameter "Q" (microns), the tap fully retracts to remove the chip. = 0 After setting each peck by using parameter "Q" (microns), the tap does not fully retract; it stops and then performs the second tapping pass, continuing in this way until it reaches the arrival coordinate.

Programming example:
G88 X26 P500 **Q5000** F1 K1

16.4 Cycle cancelation functions

The cycle can be canceled by using G80, a modal G-code that repeats the tapping operation for each position programmed after the cycle activation.

16.5 Cycle parameters

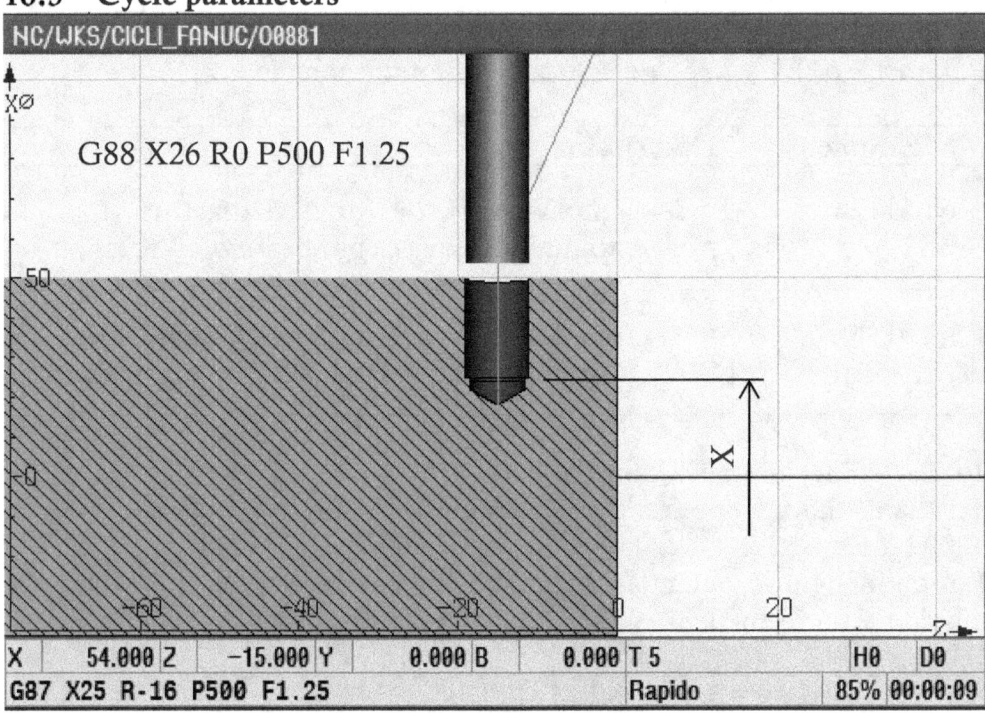

Fig. 66. G88: cycle parameters

Parameter	Description
Z	Z-coordinate of the cycle start point. If it is not programmed, the position in Z remains the position of the cycle start point.
C	Potential angular position of the workpiece. When it is programmed in the cycle, the C-axis should be enabled before recalling the cycle. Otherwise, the workpiece should be oriented angularly before the cycle

X	Absolute X-coordinate at the end of tapping.
R	Radial distance along the X-axis from the point programmed before the cycle to the tapping start point. When it is not programmed, tapping starts from the cycle start point.
P	Dwell time at the bottom of the hole expressed in milliseconds.
F	Lead of the tap.
K	Number of cycle repetitions. This parameter needs to be programmed together with the incremental distance between the holes. Example 1: program "W-10" and "K2" to tap two holes at a 10 mm distance along the Z-axis. Example 2: program "H90" and "K4" to tap four holes at a 90° angle along the "C-axis". Program the functions for the activation of the "C-axis" before the cycle. If K0 is selected, no hole is made. After the angular orientation, some machines require brake activation to lock the spindle. Hence, program the M function according to the manufacturer's instructions (e.g.: H90 K4 M31).

16.6 Programming example

16.6.1 Radial tapping

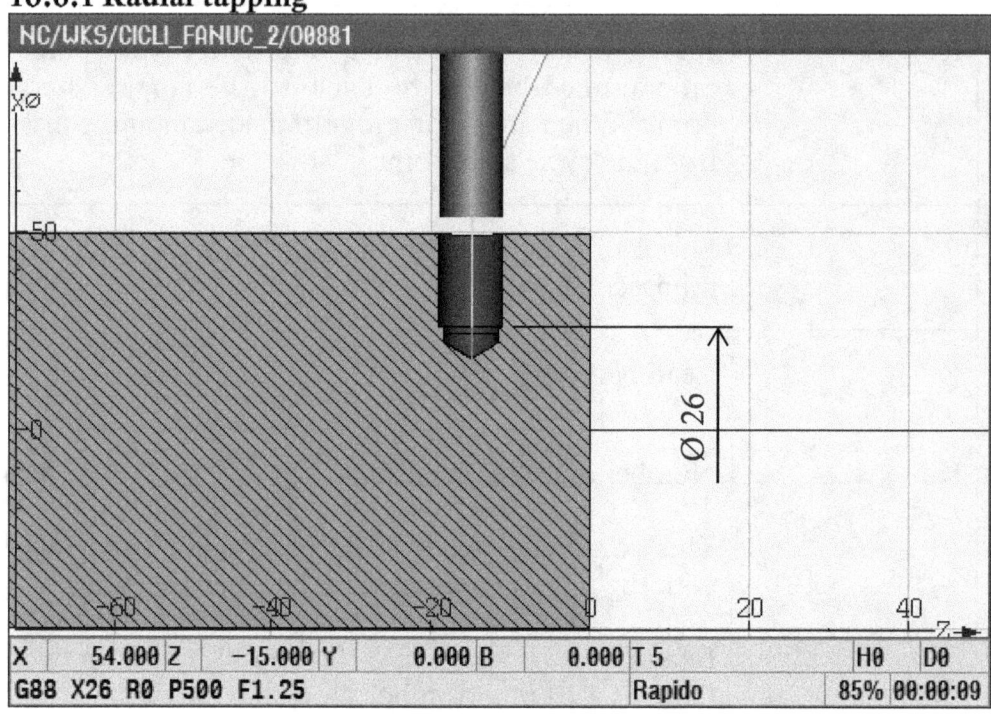

Fig. 67. G88: programming example

```
%
O0881
(G88 RADIAL TAPPING M12 F1.75)
G290 ;SIEMENS LANGUAGE ACTIVATION
WORKPIECE(,,,"CYLINDER",192,0,-90,-80,50)
G291 ;FANUC LANGUAGE ACTIVATION
G18 (X-Z PLANE)
G90 (ABSOLUTE PROGRAMMING)

T0303 (RADIAL DRILL D.6.8)
G290 ;ENABLE SIEMENS LANGUAGE TO SELECT THE DRIVEN TOOL AS
MAIN SPINDLE
SETMS(3)
G291 ;FANUC LANGUAGE ACTIVATION
G97 S1400 M3 (SPINDLE CONSTANT REVOLUTION)
G95 (WORKING FEED RATE IN MM/REV)
G0 X54 Z-15 (CYCLE START POINT)

M19 B0 (MAIN SPINDLE ANGULAR ORIENTATION)
```

```
G87 X18 P500 Q4000 F0.12 K1
G80 (CYCLE CANCELATION)
G0 X100

T0505 (RADIAL TAPPING TOOL M8 F1.25)
G97 S800 M3 (SPINDLE CONSTANT REVOLUTION)
G95 (WORKING FEED RATE IN MM/REV)

G0 X54 Z-15 (CYCLE START POINT)
```
M29 S800 (RIGID TAPPING ACTIVATION)
G88 X26 R0 P500 F1.25
G80 (TAPPING DEACTIVATION)

```
G0 Z200
G0 X200

M5
M30

%
```

17. G85: boring cycle along the Z-axis
(G83-A, G83-C)

17.1 Description

This cycle performs boring or reaming operations along the Z-axis.
Unlike the drilling cycle, the boring cycle performs a working return movement.

The cycle carries the following movements.
1   The cycle starts from the point programmed before the cycle.

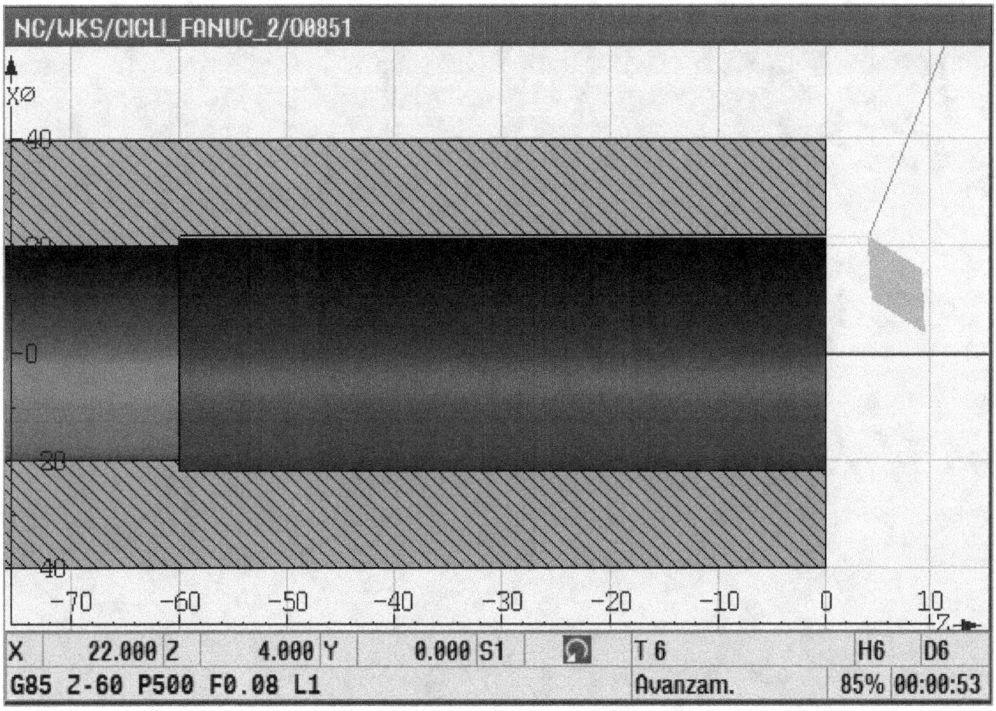

Fig. 68. G85: cycle movements

2   Rapid movement to the "X-coordinate" programmed in the cycle. This parameter defines the boring diameter. When the parameter is not programmed, boring is performed at the point set before the cycle.

3   Rapid movement to the point defined in parameter "R". When the parameter is not programmed, boring starts from the point set before the cycle.

4   Boring is performed with a single cut until the cycle reaches the "Z-coordinate" programmed in the cycle.

5   At the end of the cut, the tool is retracted from the bottom of the hole to the point defined in parameter "R" with double working feed rate value. Then, it rapidly reaches the point programmed before the cycle.

17.2 Cycle cancelation functions

Use G80 to cancel this cycle.

17.3 Cycle parameters

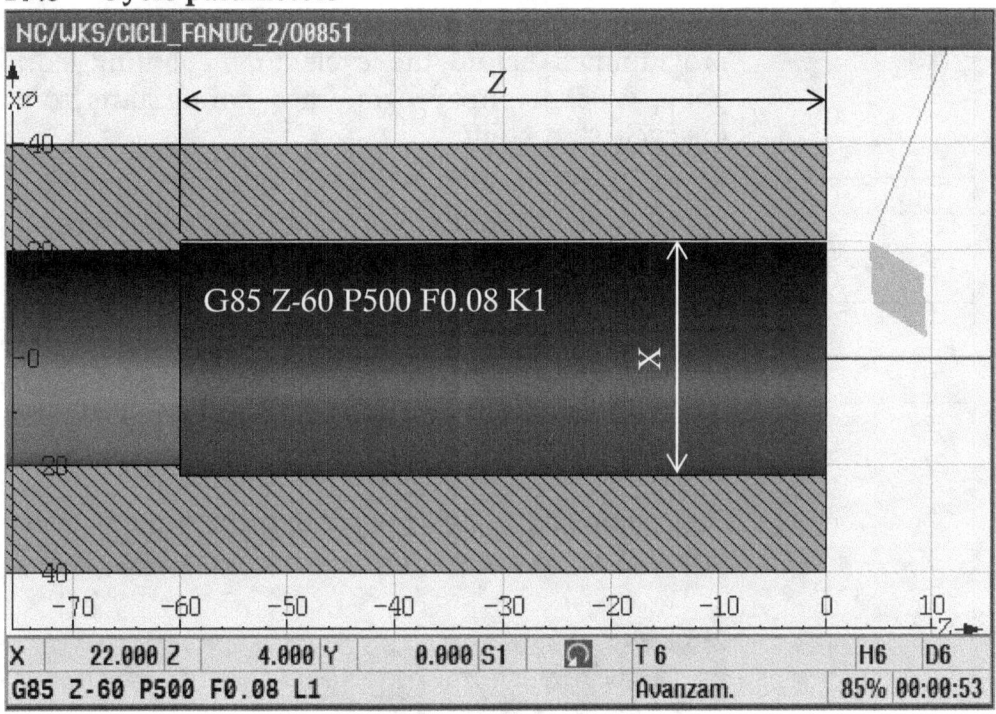

Fig. 69. G85: cycle parameters

Parameter	Description
X	X-coordinate of the cycle start point. If it is not programmed, the position in X remains the position of the point set before the cycle.
C	Potential angular position of the workpiece. When it is programmed in the cycle, the C-axis should be enabled before recalling the cycle. Otherwise, the workpiece should be oriented angularly before the cycle

Z	Absolute coordinate along the Z-axis of the cut arrival point.
R	Distance along the Z-axis from the point programmed before the cycle to the boring start point. When it is not programmed, boring starts from the cycle start point.
P	Dwell time at the bottom of the hole (expressed in milliseconds).
F	Working feed rate.
K	Number of cycle repetitions. This parameter is used to bore or ream multiple front holes. Program the functions for the activation of the "C-axis" before the cycle.

	This parameter needs to be programmed together with the incremental distance between the holes. Program "W-10" and "K2" to tap two holes at a 10 mm distance along the Z-axis. If K0 is selected, no hole is made. After the angular orientation, some machines require brake activation to lock the spindle. Hence, program the M function according to the manufacturer's instructions (e.g.: H90 K4 M31).

17.4 Programming example

17.4.1 Axial boring

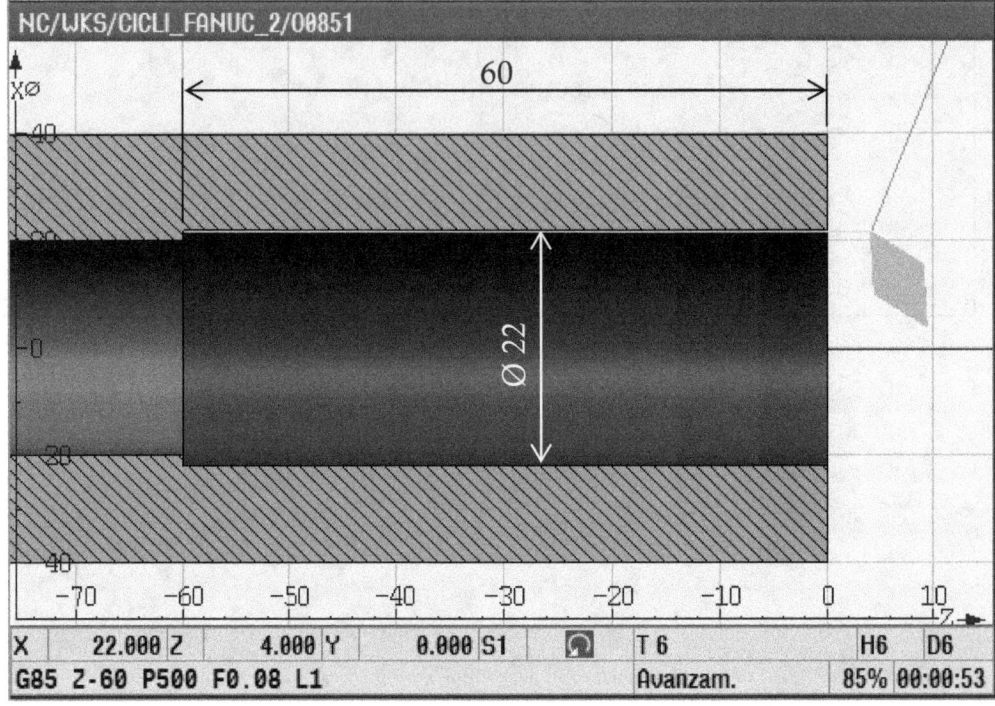

Fig. 70. G85: programming example

```
%
O0851
(G85 AXIAL BORING)
G290 ;SIEMENS LANGUAGE ACTIVATION
WORKPIECE(,,,"PIPE",448,0,-90,-80,40,20)
G291 ;FANUC LANGUAGE ACTIVATION
G18 (X-Z PLANE)
G90 (ABSOLUTE PROGRAMMING)
T0606 (BORING BAR)
G97 S1400 M4 (SPINDLE CONSTANT REVOLUTION)
G95 (WORKING FEED RATE IN MM/REV)

G0 X22 Z4 (CYCLE START POINT)
G85 Z-60 P500 F0.08 K1
G80 (CYCLE CANCELATION)
G0 X200 Z200
M5
M30
%
```

18. G89: boring cycle along the X-axis
(G89-A, G89-C)

18.1 Description
This cycle performs boring or reaming operations along the X-axis.
Unlike the drilling cycle, the boring cycle performs a working return movement.

The cycle carries out the following movements.
1. The cycle starts from the point programmed before the cycle.

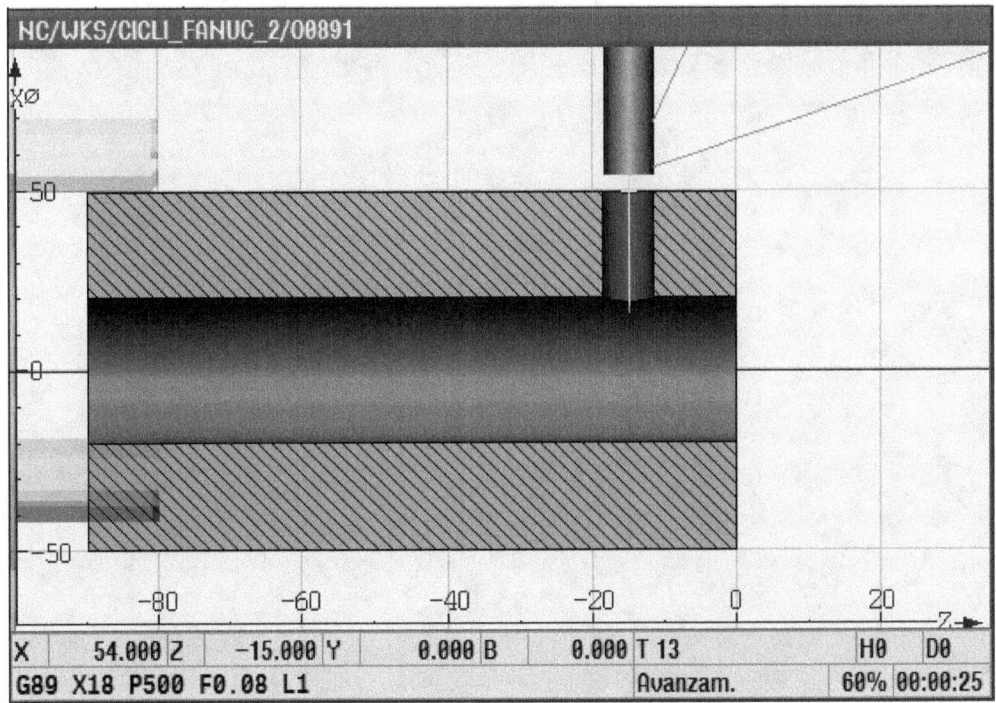

Fig. 71. G89: cycle movements

2. Rapid movement to the "Z-coordinate" programmed in the cycle. This parameter defines the boring position. When the parameter is not programmed, boring is performed at the point set before the cycle.

3. Rapid movement to the point defined in parameter R. When the parameter is not programmed, boring starts from the point set before the cycle.

4. Boring is performed with a single cut until the cycle reaches the "Z-coordinate" programmed in the cycle.

5. At the end of the cut, the tool is retracted from the bottom of the hole to the point defined in parameter R with double working feed rate value. Then, it rapidly reaches the point programmed before the cycle.

18.2 Cycle cancelation functions
Use G80 to cancel this cycle.

18.3 Cycle parameters

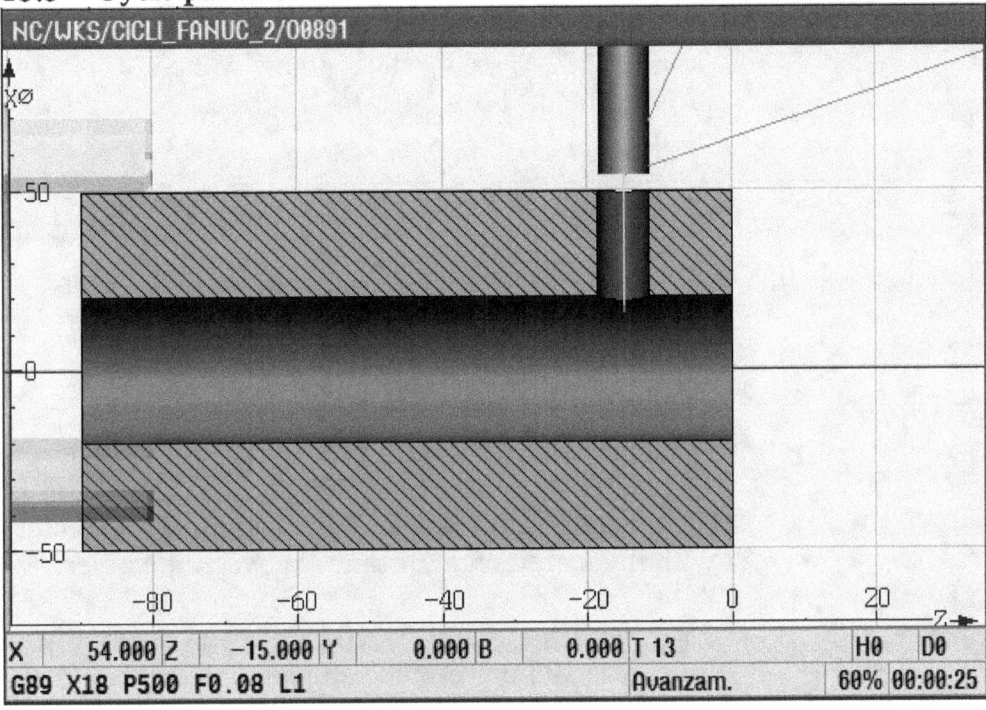

Fig. 72. G89: cycle parameters

Parameter	Description
Z	Z-coordinate of the cycle start point. If it is not programmed, the position in Z remains the position of the point set before the cycle.
C	Potential angular position of the workpiece. When it is programmed in the cycle, the C-axis should be enabled before recalling the cycle. Otherwise, the workpiece should be oriented angularly before the cycle

X	Absolute coordinate along the X-axis of the cut arrival point.
R	Radial distance along the X-axis from the cycle start point to the boring start point. When it is not programmed, boring starts from the cycle start point.
P	Dwell time at the bottom of the hole (expressed in milliseconds).
F	Working feed rate.
K	Number of cycle repetitions. This parameter needs to be programmed together with the incremental distance between the holes. Example 1: program "W-10" and "K2" to bore two holes at a 10 mm distance along the Z-axis. Example 2: program "H90" and "K4" to bore four holes at a 90° angle along the "C-axis". Program the functions for the activation of the "C-axis" before the cycle. If K0 is selected, no hole is made. After the angular orientation, some machines require brake activation to lock the spindle. Hence, program the M function according to the manufacturer's instructions (e.g.: H90 K4 M31).

18.4 Programming example

18.4.1 Radial reaming

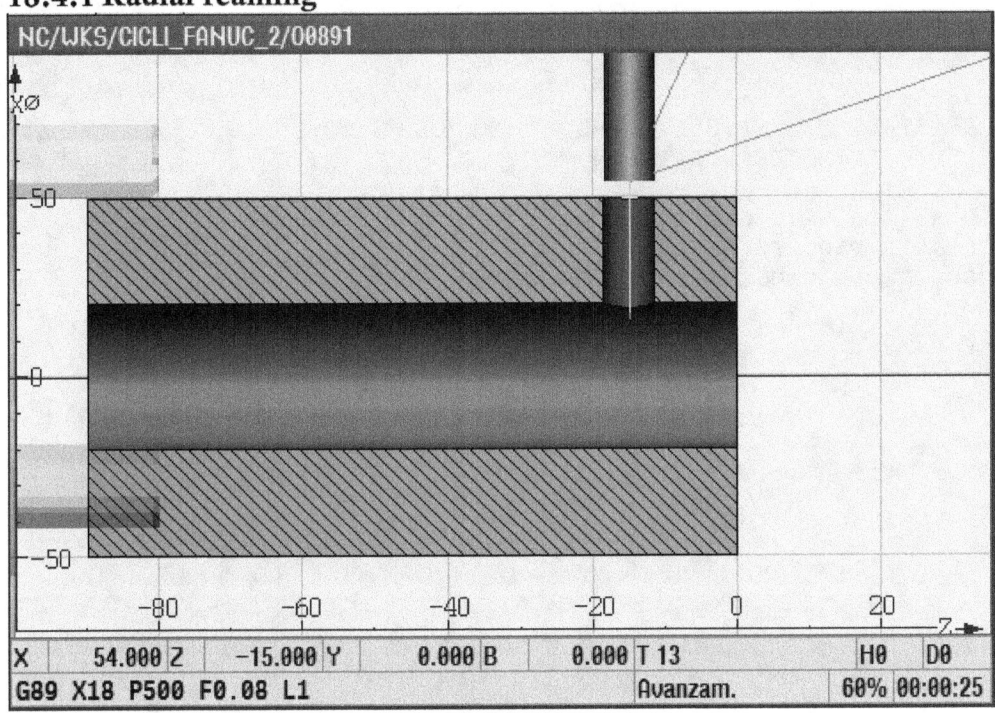

Fig. 73. G89: programming example

```
%
O0891
(G89 RADIAL REAMING)
G290 ;SIEMENS LANGUAGE ACTIVATION
WORKPIECE(,,,"PIPE",448,0,-90,-80,50,20)
G291 ;FANUC LANGUAGE ACTIVATION
G18 (X-Z PLANE)
G90 (ABSOLUTE PROGRAMMING)

T0303 (RADIAL DRILL D.6.8)
G290 ;ENABLE SIEMENS LANGUAGE TO SELECT THE DRIVEN TOOL AS
MAIN SPINDLE
SETMS(3)
G291 ;FANUC LANGUAGE ACTIVATION
G97 S1400 M3 (SPINDLE CONSTANT REVOLUTION)
G95 (WORKING FEED RATE IN MM/REV)
G0 X54 Z-15 (CYCLE START POINT)

M19 B0 (MAIN SPINDLE ANGULAR ORIENTATION)
```

```
G87 X16 P500 Q4000 F0.12 K1
G80 (CYCLE CANCELATION)

G0 X200 Z200

T1313 (RADIAL REAMER D.6.9)
G97 S1400 M3 (SPINDLE CONSTANT REVOLUTION)
G95 (WORKING FEED RATE IN MM/REV)

G0 X54 Z-15 (CYCLE START POINT)
G89 X18 P500 F0.08 L1
G80 (CYCLE CANCELATION)

G0 X200
M5
M30
%
```

148

www.ingramcontent.com/pod-product-compliance
Lightning Source LLC
Chambersburg PA
CBHW062324220526
45469CB00008B/2610